George Samuel Newth

Elementary Inorganic Chemistry

George Samuel Newth

Elementary Inorganic Chemistry

ISBN/EAN: 9783743343580

Manufactured in Europe, USA, Canada, Australia, Japa

Cover: Foto ©berggeist007 / pixelio.de

Manufactured and distributed by brebook publishing software (www.brebook.com)

George Samuel Newth

Elementary Inorganic Chemistry

ELEMENTARY CHEMISTRY

ELEMENTARY
INORGANIC CHEMISTRY

BY

G. S. NEWTH, F.I.C., F.C.S.

DEMONSTRATOR IN THE ROYAL COLLEGE OF SCIENCE, LONDON
ASSISTANT EXAMINER IN CHEMISTRY, SCIENCE AND ART DEPARTMENT

AUTHOR OF

"A TEXT-BOOK OF INORGANIC CHEMISTRY," "CHEMICAL
LECTURE EXPERIMENTS," ETC.

NEW YORK
LONGMANS, GREEN, AND CO.
LONDON AND BOMBAY

1899

PREFACE

THIS book has been written to meet the modern and practical methods of science teaching which are now being universally recognized and adopted in schools and colleges.

Formerly students were taught chemistry in the lecture-room, the knowledge so gained being supplemented by a minimum amount of practical work, and that almost exclusively analytical. The tendency of the present day is to make the student, from the very beginning, an *investigator ;* to train and develop his faculties for observation ; to make him find out facts and discover truths for himself ; in other words, to make him *think* instead of merely committing to memory what others have thought. I have therefore endeavoured, as far as it is possible to do so in a text-book, to fall into line with these views. In actual practice the *purely* inductive method of instruction breaks down. There is so much that the student is required to learn, that life itself is not long enough, and certainly the limited time at the disposal of the student is all too short, to admit of his going through the necessarily slow process of gaining this knowledge by his own investigations. *Some* facts he must take on trust, and the question therefore resolves itself into the judicious selection on the part of the teacher of the facts he will endeavour to let his students find out for themselves, and those he will teach them, and expect them to commit to memory.

In a text-book it is almost inevitable that, in giving such directions as will lead a student on to the discovery of a fact, the fact itself shall be stated.

Before introducing the student to the study of any of the elements, I have sought to familiarize him with a number of important common laboratory processes, in a chapter on " Simple Manipulations ; " and this is followed by short chapters on the " Fitting up of Apparatus," and " Simple Glass-blowing Operations."

After hydrogen, oxygen, and water have been studied, I have introduced, under the head of "Simple Quantitative Manipulations," a number of experiments or exercises involving the operations of weighing and measuring. These experiments have been selected with the object of leading the student on to the discovery of some of the fundamental laws of chemistry, making use of such knowledge of chemical facts as he has already gained. In order that he may do these with an entirely unprejudiced or unbiassed mind, they have intentionally been introduced *before* he has learnt the use of symbols and formulæ, or how to calculate what results he *ought* to get. For some of these experiments I am indebted to the suggestions of Dr. Tilden, made during the course of a short series of lectures to science teachers, at the Royal College of Science, London, in July, 1895.[1]

---

[1] Since embodied in a little publication, "Hints on the Teaching of Elementary Chemistry" (Longmans, Green, and Co.).

In an elementary practical text-book it would obviously be unwise to take up the study of more than a very limited number of the elements and their compounds. Exactly which elements and which of their compounds are the most suitable for the purpose is probably a point on which teachers will hold different opinions. I have selected those which, in my judgment, are the best adapted for an elementary course, and which I consider are well calculated to give the student a broad, general foundation upon which he can afterwards build.

G. S. N.

TABLE OF CONTENTS

LIST OF ELEMENTS

Names.	Atomic Symbols.	Atomic Weights.	Names.	Atomic Symbols.	Atomic Weights.
Aluminium . .	Al	27	*Molybdenum* .	Mo	96
Antimony (*Stibium*)	Sb	120	Nickel . .	Ni	59
Argon . .	A	(?)	*Niobrum* . .	Nb	93·7
Arsenic . .	As	75	Nitrogen . .	N	14
Barium . .	Ba	137	*Osmium* . .	Os	191
Beryllium [1] . .	Be	9	Oxygen . .	O	16
Bismuth . .	Bi	207·5	*Palladium* .	Pd	106
Boron . .	B	11	Phosphorus .	P	31
Bromine . .	Br	80	Platinum . .	Pt	195
Cadmium . .	Cd	112	Potassium (*Ka-*		
Cæsium . .	Cs	133	*lium*) . .	K	39
Calcium . .	Ca	40	*Rhodium* . .	Rh	104
Carbon . .	C	12	*Rubidium* .	Rb	85
Cerium . .	Ce	141	*Ruthenium* .	Ru	103·5
Chlorine . .	Cl	35·5	*Samarium* .	Sm	150
Chromium . .	Cr	52	*Scandium* .	Sc	44
Cobalt . .	Co	59	Selenium . .	Se	79
Copper (*Cuprum*)	Cu	63	Silicon . .	Si	28
Didymium . .	Di	145	Silver (*Argentum*)	Ag	108
Erbium . .	Er	166	Sodium(*Natrum*)	Na	23
Fluorine . .	F	19	Strontium .	Sr	87·3
Gallium . .	Ga	70	Sulphur . .	S	32
Germanium. .	Ge	72	*Tantalum* .	Ta	182
Gold (*Aurum*) .	Au	197	*Tellurium* .	Te	125
Hydrogen . .	H	1	*Thallium*. .	Tl	203·7
Indium . .	In	113	*Thorium* . .	Th	232
Iodine . .	I	127	Tin (*Stannum*) .	Sn	118
Iridium . .	Ir	192·5	*Titanium* .	Ti	48
Iron (*Ferrum*) .	Fe	56	*Tungsten* . .	W	184
Lanthanum. .	La	138·5	*Uranium* . .	U	239·8
Lead (*Plumbum*) .	Pb	207	*Vanadium* .	V	51·1
Lithium . .	Li	7	*Ytterbium* .	Yb	173
Magnesium . .	Mg	24	*Yttrium* . .	Y	89·6
Manganese . .	Mn	55	Zinc . .	Zn	65
Mercury (*Hydrar-*			*Zirconium* .	Zr	90·4
gyrum) . .	Hg	200			

[1] Those printed in italics may be regarded as rare substances.

ELEMENTARY

PRACTICAL CHEMISTRY.

CHAPTER I.

STATES OF MATTER—PHYSICAL AND CHEMICAL CHANGE.

ALL material things with which we are acquainted exist under ordinary circumstances in one of three states or conditions—either they are *solids*, like chalk, iron, sulphur, ice; or *liquids*, as water, alcohol, mercury; or they are *gases*, like air, oxygen, and steam.

We know many substances, however, which can easily be made to pass from one of these states to the other; thus it is familiar to all that when *solid water* (that is, *ice*) is gently warmed it changes into *liquid water*, and that when this liquid is more strongly heated it passes off into steam, which is water in the gaseous or vaporous state. We also know that when steam is cooled it changes back again to liquid water, and that when this is further cooled it is turned into ice, or solid water.

Sulphur is an example of another common substance that can readily be caused to change from one state to another.

Experiment 1.—Take a piece of sulphur (*brimstone*) and chip off a fragment about the size of a pea. (Note that sulphur is a pale yellow solid, easily broken, being very brittle.) Place the small piece in a clean dry test-tube, and gently heat it over a small Bunsen flame, in the manner shown in the figure. Notice that the solid quickly melts, and, if *carefully* heated, is changed into a pale yellow liquid.   Now heat more strongly, and observe that the colour rapidly darkens, and the liquid presently boils.   It is now being changed from the liquid to the gaseous state.   As the gaseous sulphur reaches the part of the tube a little removed from the flame, it soon becomes cooled again, and consequently changes back again to the liquid state.   Notice liquid sulphur collecting on the upper part of the tube in the form of small yellow drops.   Continue heating until the whole of the original fragment of sulphur has disappeared from the bottom of the test-tube. Allow the tube to cool, and observe that the yellow liquid which had condensed upon the upper part gradually changes back to the *solid* state.   In this experiment, therefore, solid sulphur has been changed first to the liquid and then to the gaseous state ; and gaseous sulphur has been allowed to pass back again, first to the liquid and then to the solid condition.

F G. I.

Experiment 2.—Gently heat a small quantity of alcohol or methylated spirit in a test-tube fitted with a cork and bent glass tube (as shown in Fig. 2), which is joined to a second test-tube provided with a cork with two bent tubes.   This second tube is placed in a glass containing cold water, to keep it cool.   The spirit soon begins to boil, and to pass from the liquid to the gaseous, or vaporous state.   As this gaseous alcohol passes into the cooled tube, however, it again returns to the liquid condition, and will be seen collecting at the bottom of the tube.

We see, from these examples, that with some things it is simply a question of whether they are heated or whether they are cooled that decides the particular *state* they shall be in. This is also true of a number of other substances ; for instance, the metal mercury (*quicksilver*) is familiar to us as a *liquid*, but if we happened to be living in the Arctic regions, we should know it as a hard *solid*, resembling lead.   Again, spectrum analysis has taught us that iron, lead, copper, tin, and many

other metals which are familiar to us as solids, are present in
the sun, and that, owing to the intense heat, they exist there
in the *gaseous* state, forming a part of the sun's atmosphere.
It is not necessary to take these metals to the sun, however,
to make them change their state from the solid to the gaseous.
When we apply heat to them, some of them, such as lead and
tin, melt readily enough ; others, like copper and iron, require
a much higher temperature to make them pass into the liquid
state ; while all of them can be boiled and made to pass into
the gaseous condition by means of the electric furnace.

All solids and liquids are visible and tangible ; gases, on

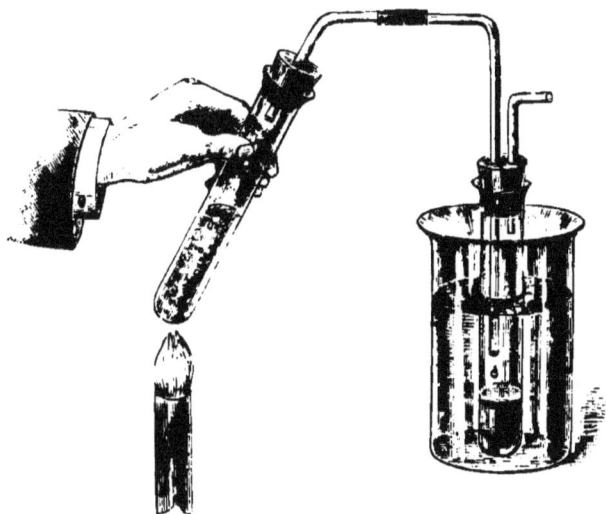

Fig. 2.

the other hand, cannot be *felt,* and in most cases they are
invisible. It is quite impossible, by merely looking at it, to
tell whether a glass bottle is filled with air, or hydrogen, or
oxygen, or whether it is entirely empty—that is, vacuous ; hence,
when a liquid substance passes into the gaseous state, it usually
disappears altogether from our sight. For example, if a
small quantity cf water be left exposed in a shallow dish or
saucer, we know that it gradually diminishes in quantity, and
finally disappears entirely. In common language, we say that
it has *dried up;* in more scientific phraseology, we speak of the
process as *evaporation*—that is to say, the water has passed

from the liquid to the vaporous or gaseous state, in which condition it is invisible, and mingles with the other equally invisible gases of the air. The visible cloud which appears when steam is allowed to escape from a boiler or locomotive engine, and which is popularly called "steam," consists in reality of minute drops of *liquid* water, and is not water in the vaporous or gaseous state. The true steam is invisible, but on suddenly coming into contact with cool air the gaseous water changes to liquid water and becomes visible. That gaseous water is invisible may be proved by the following experiment :—

Experiment 3.—Place a small quantity of water in a large glass flask, which is provided with a cork carrying a short glass tube bent at right angles. Boil the water rapidly, until a jet of steam escapes from the tube and produces the familiar cloud of "steam." It is evident that the flask must now be full of gaseous water, but it is perfectly clear and invisible. Now hold immediately beneath the jet of steam another Bunsen flame (as at 2, Fig. 3), and observe that the cloud instantly disappears. Although the same amount of steam is escaping from the tube, it is now invisible. This is because the flame warms the air in the immediate neighbourhood of the jet, so that the gaseous water does not become suddenly cooled on issuing into the air, and therefore does not condense to the liquid state.

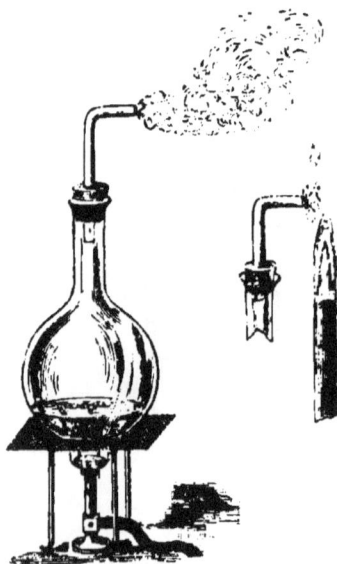

FIG. 3.

Some solid forms of matter, when heated, change into the gaseous state without first becoming liquid; that is to say, they pass at once into gases without melting. For example—

Experiment 4.—Heat in a dry test-tube a fragment of ammonium chloride (*sal ammoniac*) about the size of a grain of wheat.

Notice that the white solid does not melt, but at once passes into vapour. The vapour, on reaching the cooler parts of the test-tube, changes back again to the solid state, and collects as a white coating upon the glass. This process is termed *sublimation ;* the ammonium chloride is said to *sublime* when heated. Any impurities present in the original solid, which do not change into vapour at the temperature employed, will obviously be left behind, hence this process may be used to purify substances like ammonium chloride.

Many kinds of matter, when experimented upon in the same way as the sulphur (Exp. 1) and the ammonium chloride (Exp. 4), undergo a different kind of change, a change which is more subtle and less simple. This will be seen by the following examples :—

Experiment 5.—Heat in a dry test-tube a small quantity of potassium chlorate, and carefully notice what takes place. The solid quickly melts and changes into the liquid state, and presently appears to be boiling. So far it seems to behave like the sulphur (Exp. 1). But is the liquid potassium chlorate being changed into the gaseous state? In the first place, it will be noticed that practically nothing condenses upon the upper and cooler part of the tube ; this seems to imply that the substance is not passing into the gaseous condition. Apply a lighted taper to the mouth of the tube ; no gas is escaping which will take fire, but notice that the flame of the taper becomes brighter. Dip into the test-tube a match or splinter of wood which has been lighted and blown out, and has still a glowing spark upon it ; the wood will be rekindled and burst into flame. This proves that *something* gaseous is coming from the boiling liquid. Now fit a cork and bent tube into the test-tube, and connect it to a second tube arranged as in Fig. 2. It will be found that nothing visible, either liquid or solid, collects in the cold tube. As the heating of the potassium chlorate is continued, it will be seen that the liquid becomes less fluid, and finally goes solid again. By this experiment, therefore, we find that when solid potassium chlorate is heated it first becomes liquid, and then is changed into two different things, namely, a colourless gas which does not change either to a liquid or a solid when cooled by cold water, and a solid which is evidently different from the original one, because it is very much more difficult to melt.

Experiment 6.—Place a few grains of mercuric oxide in a test-tube, and apply heat. In this case the red solid becomes darker in colour, but does not melt. Gradually, however, there will collect upon the cooler part of the tube a *sublimate* which has the appearance of a bright white silvery metal, and upon dipping into the test-tube a glowing splint of wood, we shall obtain the same result as in Exp. 5. By continuing the heat, the whole of the original substance will disappear. We learn from this experiment, therefore, that when mercuric oxide is heated, it also changes into two different things ; one of them being a colourless gas which passes away (presumably the same gas as was given out by the potassium chlorate in Exp. 5), while the other is a white metal. Examine this metallic sublimate carefully, and notice that it consists of minute globules of liquid metal. This must be mercury, because no other metal is liquid at the ordinary temperature.

Certain differences between the kind of change undergone by the mercuric oxide and the potassium chlorate, and that experienced by sulphur and by ice, when these things are heated, will be evident. (1) Neither the sulphur nor the ice change into *two* different kinds of matter at once. (2) When sulphur or ice are changed by heat into the liquid or gaseous state, the change is only a *temporary* one ; when cooled again, they change back to their original condition. On the other hand, the mercuric oxide and the potassium chlorate each change into two different forms of matter, and the change in each case is permanent. Changes like those experienced by mercuric oxide and potassium chlorate are called **chemical changes**, whilst those which the ice and the sulphur underwent are distinguished as **physical changes**. Many chemical changes are constantly going on around us in the familiar processes of everyday life : thus, when a candle burns, the solid wax is transformed into certain invisible gases which mix with the air, and never return again to the original condition of the wax ; the candle undergoes a chemical change. In our ordinary fires, coal is converted into smoke and certain invisible gases, which escape into and pollute the atmosphere, and a small residue of a greyish ash is left ; the change is permanent, and is a chemical change.

When an egg is cooked, the clear and almost colourless liquid albumen (*white of egg*) is converted into a white solid, which does not again return to the liquid state on cooling ; the albumen undergoes a chemical change.

Human beings eat bread, meat, vegetables, etc. ; these foods undergo chemical changes, whereby they are converted into flesh and bones, into invisible gases which leave the body in the breath, and into waste products which leave the body in the perspiration and evacuations.

Chemical and physical changes, however, are very closely associated, and although we may have a physical change *without* any chemical change, all chemical changes are accompanied by a physical change ; and in many cases the accompanying physical change is the only outward indication which we have that a chemical change has taken place at all. For example—

Experiment 7.—Take a glass tube about two feet long and half an inch wide, and close one end like the bottom of a test-tube. Fit a cork into the other end, and slip an indiarubber ring upon the tube a short distance from the corked end. About half fill the tube with strong sulphuric acid, and then gently fill the tube up to the ring with cold water. The water, being much lighter than the acid, will float upon it without mixing, if poured in gently. Now tightly cork the tube, and tip it up and down two or three times in order to mix the contents together. A chemical change takes place, but nothing is to be *seen :* it will soon be found, however, that the tube is getting so hot that it can scarcely be held in the hand. Now hold the tube in an upright position, and notice that the liquid has *shrunk*, for it no longer reaches to the mark upon the tube. Here we see the chemical change is accompanied by a change of *temperature*, and a change of *volume*.

CHAPTER II

Elements and Compounds.—When different kinds of matter are experimented upon, it is found that from some of them it is possible to obtain two or more substances totally unlike the original matter, while from others it is impossible to obtain anything essentially different by any process at present known. For example, in Exp. 5 we found that from potassium chlorate we were able to obtain a gas which caused a glowing splint of wood to re-light (a gas called *oxygen*), and also a white solid residue which required a much stronger heat to melt it than the original substance did. Again, in Exp. 6, from the red mercuric oxide we obtained the same gas oxygen, and a quantity of metallic mercury, two things entirely different from the original. Substances like potassium chlorate and mercuric oxide are called *compounds*. Substances from which we are unable to obtain anything essentially different, are distinguished as *elements*.

There are only about 70 substances which are believed to be elementary bodies; and nearly the half of these may be considered as rareties. The following list of thirty includes all the most important of the elements (for the complete list see inside the cover) :—

Aluminium, Al.	Calcium, Ca.	Gold, Au.
Antimony, Sb.	Carbon, C.	Hydrogen, H.
Arsenic, As.	Chlorine, Cl.	Iodine, I.
Bismuth, Bi	Copper, Cu.	Iron, Fe.
Bromine, Br.	Fluorine, F.	Lead, Pb.

Magnesium, Mg.	Oxygen, O.	Silver, Ag.
Manganese, Mn.	Phosphorus, P.	Sodium, Na.
Mercury, Hg.	Platinum, Pt.	Sulphur, S.
Nickel, Ni.	Potassium, K.	Tin, Sn.
Nitrogen, N.	Silicon, Si.	Zinc, Zn.

Of these elements two are liquid at the ordinary temperature, namely *bromine* and *mercury ;* five are gases, *chlorine, fluorine, hydrogen, nitrogen,* and *oxygen,* while the rest are solid. Most of the solid elements are metals, and the above list will be seen to contain the names of such familiar metals as copper, gold, iron, lead, etc. Those of the solid elements in this list that are not metals are arsenic, carbon, iodine, phosphorus, silicon, and sulphur. Some of these possess, in a greater or less degree, some of those properties which are usually associated with the metals. Thus arsenic and silicon are opaque substances having the power of reflecting light from their surfaces, a property usually known as *metallic lustre.* Carbon, again, has the power of conducting both heat and electricity, properties which are almost exclusively characteristic of metals. The name *metalloids* is sometimes given to those elements which, while not being true metals, yet closely resemble them in some of their properties.

Mechanical Mixtures.—When two elements are brought together, either nothing happens, or else a chemical change takes place. In the former case the result is a simple mixture of the two elements ; in the latter it is the formation of a compound. Similarly, when two compounds are brought together, if a chemical change takes place it results in the formation of new compounds, whereas if no chemical change follows we only obtain a mechanical mixture of the two compounds.

Experiment 8.—Grind to powder in a mortar a small quantity of potassium chlorate, and mix it with about the same quantity of powdered white sugar. No chemical change takes place, therefore the result is a *simple mixture.*

Experiment 9.—Place in a dry test-tube a fragment of phosphorus about the size of a pea, and drop upon it a few grains of powdered iodine. A chemical change at once takes place, great heat is evolved, and a *compound* of phosphorus with iodine results.

[Phosphorus is a substance which must be handled with great care, as it very easily takes fire. It is always preserved beneath water, and when a small fragment is required for experiment, the larger piece should be taken out of the bottle by means of tongs, placed on a plate with some water, and there cut with a penknife. The piece to be used must be quickly wiped with blotting-paper, and put into the test-tube with the tongs. It should never be taken up by the fingers, as the warmth of the hand might cause it to take fire.]

Experiment 10.—Place in a dry test-tube a few drops of mercury, together with a few particles of powdered iodine, and gently heat the tube. A chemical change follows, and a *sublimate* is obtained partly red and partly yellow. This sublimate consists of a *compound* of mercury and iodine.

Each of the ingredients in a simple mixture retains its own individual and characteristic properties. Thus, if the mixture of potassium chlorate and sugar (Exp. 8) be tasted, both the sweet taste of the sugar and the peculiar saline taste of the potassium chlorate will easily be perceived.

Owing to the fact that their properties are retained, the ingredients of a mechanical mixture can be separated from each other by processes which are purely mechanical or physical, as distinguished from chemical.

Experiment 11.—Powder some potassium nitrate (*nitre*), and mix into it a little powdered sulphur (*flowers of sulphur*) and a small quantity of fine iron filings. The result is a mechanical mixture of these things. If this mixture be now examined through a pocket lens, the separate particles of white nitre, yellow sulphur, and grey iron will be distinctly visible, lying side by side unchanged. Place the mixture upon a sheet of white paper and bring a magnet near to it ; the iron in the mixture will be attracted by the magnet, and may in this way be drawn away and entirely removed. The residue, containing now only the nitre and sulphur, should be removed to a test-tube, a little water added, and the contents of the tube warmed. The nitre being easily dissolved by water is thus separated from the sulphur, which does not dissolve. If the mixture be now poured upon a blotting-paper filter (see p. 17) the watery solution containing the nitre passes through, while the sulphur remains on the paper. If the liquid be then boiled in a little dish, the water will evaporate and the nitre will be left as a white solid residue in the dish. In this way the three ingredients have been separated by physical processes.

Chemical Affinity.—In a compound the various constituents stand in a totally different relation to each other than is the case with mixtures ; a relation which is much closer, and more difficult to understand.

The elements present in a compound are said to be *chemically combined* with each other : their union in all cases is controlled by the operation of a particular force, which is spoken of as *chemical affinity*.

Consider some of the cases of chemical combination already referred to. In Exp. 10 the two elements, mercury and iodine, enter into chemical union, and the compound formed is called mercuric iodide. If this scarlet powder be examined by the most powerful microscope it is quite impossible to see either the mercury or the iodine it contains. All the properties belonging to mercury, as well as those belonging to iodine, are gone, and the compound is endowed with an entirely new set of properties which are peculiar to itself.

Again, in Exp. 6 we learnt that mercuric oxide was a compound of mercury and oxygen. These two elements, the liquid metal *mercury* and the colourless gas *oxygen*, when united together by the influence of *chemical affinity*, entirely lose their own individuality, and take on altogether new habits —the compound is a *brick-red crystalline solid*, it possesses none of the properties of either mercury or oxygen, and these constituents cannot be separated again by any purely mechanical operations.

The familiar substance water, is a compound obtained by the chemical union of two colourless invisible gases ; one of them (hydrogen) a gas which easily burns, and the other (oxygen) a gas which causes ordinary burning things to burn more quickly and brilliantly. The compound, therefore, which these gases give rise to when they chemically combine has none of the properties of the ingredients, its properties are diametrically opposed to theirs—it is liquid, it does not burn, and it is the great antidote for fire. By no mechanical method can we disengage or separate these constituents.

Why it is that the compound produced, when two such substances as hydrogen and oxygen enter into chemical union, should have the particular properties with which water is endowed, no one knows. Or why the compound of mercury and iodine should happen to be *red*, and not blue or green, we cannot tell. And, in the same way, it is quite impossible, from a knowledge of the properties of the ingredients, to foretell what will be the nature of the compound they will give rise to. For instance, given a certain gas (*chlorine*), with a greenish-yellow colour, a powerful irritating smell, and extremely poisonous; also a soft white metal (*sodium*), which if placed upon the tongue would take fire; now what sort of properties are likely to be possessed by a compound formed by the chemical union of these substances? No one would predict that the product would be the innocent and necessary article of food, *common salt;* but such is actually the case.

Chemical Action is the term applied to the actual processes which take place by the operation of the force called chemical affinity. Thus, when phosphorus unites with iodine (Exp. 9), or when iodine combines with mercury (Exp. 10), the process of combination is termed *chemical action*, we say that a chemical action takes place between the phosphorus, or the mercury, and the iodine.

Chemical action does not take place promiscuously between all the elements. Some are made to combine only with great difficulty; others will not combine together at all. As a general rule, those elements which are most *unlike* each other combine together most readily, they are said to have the greatest *affinity* for each other. Neither does chemical action take place in all cases under the same *conditions;* thus sometimes it takes place at once on simply bringing the substances together.

Experiment 12.—Place a small quantity of sodium peroxide in a dry test-tube, and pour upon the powder a little cold water.

Chemical action at once takes place, the mixture effervesces briskly. Such effervescence signifies that a gas is being set free, as one of the products of the chemical action. If a glowing match or a splint of wood be dipped into the test-tube, the same result will follow as in Exps. 5 and 6, showing that the gas given off is oxygen. See also Exp. 9.

In other cases it is necessary to employ some external energy in order to induce chemical action to begin. In a large number of instances the application of heat will bring about chemical action. Thus, in Exp. 10, chemical action between the mercury and iodine was induced by heating the mixture. This may also be exemplified by the following experiments.

Experiment 13.—Make a small heap of the mechanical mixture of potassium chlorate and sugar (Exp. 8), and apply a lighted match to it. Chemical action is at once set up, and rapidly propagated throughout the heap.

Experiment 14.—Place a few fragments of copper foil or wire in a test-tube, and pour upon them a small quantity of strong sulphuric acid. While cold no action takes place, but if the mixture be warmed it will first become muddy, and presently give off a gas (*sulphur dioxide*), which has a choking smell like that produced by burning sulphur.

Sometimes chemical action is brought about by the influence of light.

Experiment 15.—Brush over one side of a half-sheet of note-paper with a dilute solution of silver nitrate, using a clean soft brush, and put the paper away in a drawer until dry. Cut out of a piece of brown paper some figure, either a capital letter, or other design. Lay the prepared paper upon a piece of wood and fasten the brown paper figure down upon it with one or two drawing-pins ; expose it to bright sunshine for a short time. Where the prepared paper has been exposed, the sunlight will cause a chemical change to take place, which will result in the discolouration of the paper, so that on removing the brown paper, an image of the design will be seen on the under paper. This experiment is a primitive photographic process. All photography depends upon the influence of light in promoting chemical action.

In a number of cases, chemical action is only able to

proceed in the presence of a third substance, which itself remains unchanged at the conclusion of the action, and which sometimes needs only to be present in the most infinitesimal quantity. These cases are all classed together under the head of *catalytic actions*, the third substance being spoken of as the *catalytic agent*.

All known instances of chemical action take place according to one of three general modes. These will be better understood after other matters have been considered (p. 155).

CHAPTER III.

SIMPLE MANIPULATIONS.

Solution.—This term is applied both to the *act* of dissolving, and to the *product obtained* by dissolving. For example—

Experiment 16.—Throw into some water in a test-tube a little powdered potassium nitrate (*nitre*); and, on shaking, it will soon be entirely dissolved. We say that the potassium nitrate has undergone *solution*, and we term the resulting liquid a *solution* of nitre.

The liquid in which a substance is dissolved is called the *solvent*; thus, in the above example, the solvent was water. If such a solution be heated, or even allowed to stand exposed in an open dish, the solvent will gradually *evaporate* (see p. 3), and leave the dissolved substance behind.

Experiment 17.—Pour the solution of nitre from Exp. 16 into a porcelain evaporating basin, and heat gently by means of a Bunsen with a rose burner, as shown in Fig. 4. Continue the process until the water is all driven off, and a white saline residue is left in the dish. This is called *evaporating to dryness*.

FIG. 4.

Unfortunately the term *solution* is employed without distinction, to denote two essentially different processes of dissolving. This will be best understood by the following examples.

Experiment 18.—Place a small quantity of powdered sodium carbonate (*washing soda*) in two separate glasses. To one, add some water, which will dissolve the sodium carbonate, giving a clear solution. We have therefore made a solution of sodium carbonate in water. Into the second glass pour some dilute hydrochloric acid, and again the sodium carbonate dissolves, and a clear solution is obtained. In this second case, however, one striking difference will be noticed, which is, that the act of solution of the sodium carbonate in hydrochloric acid is attended by a brisk effervescence. This means that some *gas* is being disengaged, and that a *chemical change* is taking place. Now place each solution in a porcelain dish and gently evaporate to dryness. In both dishes a white residue will be left, but let us try and find out whether they are the same things or not. One property of sodium carbonate has been exhibited already during this experiment, namely, that an effervescence takes place when dilute hydrochloric acid is dropped upon it ; let us therefore use this property as a test, and add a few drops of hydrochloric acid to the residue in each dish. In the case of the residue from the watery solution, there will be an effervescence, so that we may infer that the sodium carbonate did not undergo any permanent change by being dissolved in water, and is left unaltered when the water is evaporated away. In the other case no effervescence will take place, showing that this residue is something quite different from the original substance.

Experiment 19.—Place a few scraps of copper in a test-tube, and add water ; the metal does not dissolve. Pour away the water, and add a little strong nitric acid ; effervescence takes place, a brown-coloured offensive smelling gas makes its appearance, and the copper disappears. We say, therefore, that we have made a solution of copper in nitric acid ; or that copper is soluble in nitric acid. It will be noticed that the solution has a blue colour, and if it be evaporated gently to dryness a blue residue will be left, which evidently is not *copper*. It is a compound of copper, namely, copper nitrate.

It will be evident from these experiments that two kinds of solution are recognized—one being more of a physical character, in which the substance is dissolved without apparent change, and from which it is again deposited in its original state by evaporating the solvent ; while, in the other case, the act involves a chemical change, a chemical action taking place between the solvent and the dissolved substance, giving rise to new compounds entirely different from either of these things. This latter process is sometimes distinguished as *chemical solution.*

The process of solution affords an important method for separating substances that are mixed together. Thus—

Experiment 20.—Powder a piece of marble, and mix it with powdered sodium carbonate, and throw a little of the mixture into some water in a test-tube, and shake it up. The sodium carbonate dissolves, but the insoluble marble settles to the bottom. In order to separate the solution from the sediment, one of two methods may be used, namely, *decantation* or *filtration.*

Experiment 21.—**Decantation.** Allow the marble to settle, and carefully pour off as much of the clear liquid as possible without disturbing the sediment. Nearly fill the test-tube again with water and, after thoroughly shaking, allow the marble to settle once more, and again decant the clear liquid. By repeating this process the marble will be washed free from the sodium carbonate.

Experiment 22.—**Filtration.** The filtering medium, almost exclusively used, is a bibulous paper (like blotting-paper), known as *filter-paper.* This is usually obtained already cut in circular pieces of various sizes. One of these pieces is made into a cone by being folded first into half, and then at right angles into half again, and is supported in a glass funnel of such a size that the glass will project slightly *above* the paper (Fig. 5). The cone is then wetted with a little clean water, and the filter is ready for use. The solution to be filtered is poured into the cone (without overflowing the paper), and the in-

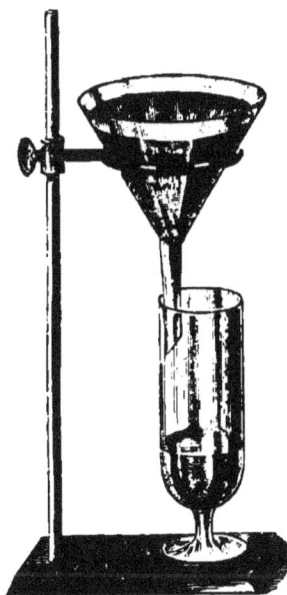

FIG. 5.

soluble matter is arrested by the paper, while the solution passes through quite clear. When the *whole* has run through, the insoluble portion is washed free from any adhering solution by once or twice filling the filter with water, and each time allowing the whole of the wash water to drain through.

By conducting such an operation carefully, the exact

quantities of both the marble and the sodium carbonate present in the mixture can be ascertained. For this purpose the insoluble residue upon the filter must be dried and weighed ; and the *filtrate*, together with all the *wash waters*, must be evaporated to dryness, and the residue (consisting of the sodium carbonate) also weighed. It will be obvious that there must be no loss of any of the solution during the whole operation. When pouring a liquid from a narrow vessel like a test-tube, there is no risk of it spilling by running down the outside of the tube ; but in the case of

FIG. 6.

wide vessels it is very likely to do so, when loss of the solution would arise. This risk is obviated by the device of pouring the liquid down against a glass rod held lightly against the edge of the vessel. Figs. 6 and 7 show the two methods. Again, in order to prevent splashing from the drops of liquid falling from the point of the funnel, the latter should be made to touch the side of the vessel placed below, so that the drops may run down the surface of the glass. To facilitate this, the stems of funnels are usually cut diagonally, as seen in Fig. 5.

FIG. 7.

In the example above given, the separation was made by the *physical* process of solution, as the sodium carbonate did not undergo any chemical change during the process ; but quite as frequently the method of *chemical* solution is employed. For example—

Experiment 23.—Mix together powdered marble and white

sand. Both substances are insoluble in water. Add to some of the mixture dilute hydrochloric acid : effervescence takes place, showing that a gas is being disengaged. When no further effervescence is noticed upon adding more of the dilute acid, the mixture may be *filtered.* The residue upon the filter is the sand. Sand is not dissolved by hydrochloric acid, but marble is. If the *filtrate* be evaporated to dryness, a residue will be obtained ; but this residue is not *marble,* because if a drop of hydrochloric acid be added to it *no* effervescence takes place. The marble and the hydrochloric acid have undergone a chemical reaction, resulting in the formation of new substances, namely, a *gas* (carbon dioxide), and the *residue* (calcium chloride).

Precipitation.—Sometimes when one clear solution is added to another, the resulting mixture is no longer clear ; something is produced which makes the liquid thick or muddy.

Experiment 24.—Dissolve a small pinch of common salt (sodium chloride) in water in a test-tube, and add a few drops of a solution of silver nitrate. Instantly the mixture becomes milky, owing to the separation of a white solid, which finally settles to the bottom of the tube. The *silver* of the silver nitrate has a strong chemical affinity for the *chlorine* of the sodium chloride, and the compound produced when these unite (namely, silver chloride) happens to be insoluble in water, and therefore the moment it is formed it separates out as a solid. This process is called *precipitation,* and the solid which is produced is termed the *precipitate.*

A precipitate may be separated from the liquid in which it is suspended, either by decantation or filtration (see p. 17).

By means of the combined processes of precipitation and filtration, it is possible to separate by chemical means substances which are mixed together in solution. For example—

Experiment 25.—Take a crystal of copper nitrate and one of silver nitrate, and dissolve them together in distilled water in a test-tube. The crystal of the copper salt will impart to the solution its own blue colour, but the liquid will be clear. Add to this a few drops of a solution of common salt. Just as in Exp. 24, there is again a white precipitate of silver chloride. Add the salt solution drop by drop until no more of the white precipitate is produced. The whole of the silver originally present in the silver nitrate is now united to chlorine, and is thrown out of the solution as insoluble silver chloride. Now filter the mixture (see p. 17), and a clear blue liquid, containing all the copper nitrate, will pass through the filter while the silver chloride collects on the filter. This method of separating is the basis of most analytical processes.

Crystallization.—When we dissolve any substance in cold water, and continue adding more of the substance, a point is ultimately reached when the water will dissolve no more. The water is then said to be *saturated* with that particular substance. If such a *cold saturated solution* be warmed, it is then able to dissolve some more of the substance, until again a point is reached when it can dissolve no more. When such a *hot saturated solution* is again cooled, that quantity of the dissolved substance which the hot solution contained, over and above the amount which the cold water could dissolve, is thrown out of solution, and in many cases it is deposited in the form of crystals.

Experiment 26.—Add powdered alum in small quantities at a time to some cold water in a test-tube, constantly shaking, and allowing each little portion to dissolve before adding more. Notice that the water gradually dissolves each additional portion of alum more slowly, until at last the solution is saturated. Now heat the solution to boiling, and notice that it will now easily dissolve a further considerable quantity of the alum. Cool the solution by dipping the test-tube into cold water, and almost immediately a quantity of alum will be thrown out of solution and deposited in the form of minute crystals. Once more warm the solution, and observe that these crystals again dissolve. Now stand the test-tube down, and allow it gradually to cool by itself, and then notice that again crystals have been deposited, but that they are much larger, so that their particular shape can easily be seen. The same *quantity* is deposited in both cases, but when quickly cooled the crystals are smaller.

Different substances are soluble in water to a very different extent; thus we find that the same quantity of cold water as is just capable of dissolving eight parts of borax, will dissolve four times as much nitre, and twenty times as much

zinc sulphate (*white vitriol*). Owing to this unequal solubility of various substances, the process of crystallization is
often made use of in order to separate substances from each
other; more especially with a view to removing from one substance any admixed impurities which are *soluble*, and which,
therefore, cannot be got rid of by filtering.

Experiment 27.—Mix together about equal quantities of
powdered potassium chlorate and potassium dichromate; place
the mixture in a beaker, and pour a little boiling water upon it.
Place the beaker over a rose burner and boil the solution, and add
just enough boiling water to entirely dissolve the mixture. The
solution now has the orange-red colour of the potassium dichromate. Remove the lamp, and after a few minutes stand the beaker
in a basin of cold water, when a crop of crystals is soon deposited.
As soon as the solution is cold, decant the clear liquid and drain
the crystals. Now pour a little cold water upon the crystals in the
beaker, and again drain them. Observe that the solution first
decanted and the wash water are coloured, but that the crystals
themselves are nearly white. Rinse them once or twice more with
small quantities of cold water, and see that each rinse-water becomes less strongly coloured and the crystals become whiter. This
shows that potassium dichromate is more soluble than potassium
chlorate : and it will be evident that by repeating the process (that
is, by once more dissolving the white salt and crystallizing it again)
we can get rid of every trace of the yellow salt.

Fusion is the term applied to the process of converting
a solid substance into the liquid state by the application of
heat. Thus when ice is warmed it *enters into a state of fusion*,
or, in other words, it melts ; and when lead is heated it also
fuses or melts. In common language the term *melt* is often
incorrectly employed to denote the process of *solution*, thus
sometimes it is said that sugar *melts* in warm water. This
confusion is to be carefully avoided.

When a substance in a state of fusion is allowed to cool
and solidify, it very commonly assumes crystalline shapes ;
thus, when water passes into the solid state we obtain a
crystalline mass of ice. The shape of the crystals of ice is
readily seen by allowing a single snow-flake to fall upon the
coat sleeve, and looking at it through a pocket lens. The

magnificent fern-like crystals of ice upon a window-pane in winter are familiar to all, and very often single star-shaped crystals of great beauty are to be seen.

Chemical action often takes place between substances in a state of *fusion*, which is incapable of taking place when they are only in solution ; for example—

Experiment 28.—Dissolve a small piece of potassium hydroxide (*caustic potash*) in water, and add to the colourless solution a few grains of powdered manganese dioxide. The black powder simply falls to the bottom of the solution, and no action takes place. Now place another piece of potassium hydroxide in a dry test-tube and heat it : the solid melts to a colourless liquid. While it is in this *fused* condition add a few particles of the manganese dioxide, and notice a very different result. The liquid turns to a blue-green colour, and the black oxide of manganese has disappeared. If this be allowed to cool, it again solidifies to a green solid mass ; and when quite cold, if water be added, it dissolves, giving a green solution. A chemical action has taken place between the manganese dioxide and the *fused* potash forming a compound called potassium manganate, but the *solution* of potash was incapable of bringing about this reaction.

Distillation.—The principle of this process has already been explained (p. 2) ; but the process is more conveniently carried on by means of the apparatus in Fig. 8. The flask, *a*, known as a "*Wurtz*" flask, contains the liquid to be distilled. Its branch tube passes through a cork in the end of the *Liebig's condenser*. This consists simply of a straight tube, T, which is jacketed by a wider tube, W, through which a stream of cold water circulates, the water entering at the lower branch tube, B, and passing out at B'. In this way the inner tube is kept cold. The neck of the Wurtz flask is fitted with a cork, through which a thermometer is fixed, in order to tell the temperature at which the liquid distils.

Experiment 29.—About half fill the Wurtz flask with water, which has been made dirty by the addition of a small fragment or clay, or a little ink, and replace the cork carrying the thermometer, the bulb of which must not reach down into the water. Boil the water by means of a Bunsen lamp, and place a clean dry flask to receive the distillate. During the whole experiment a stream of cold water must be circulating through the condenser. As soon as the water in the flask begins to boil, carefully read the thermometer.

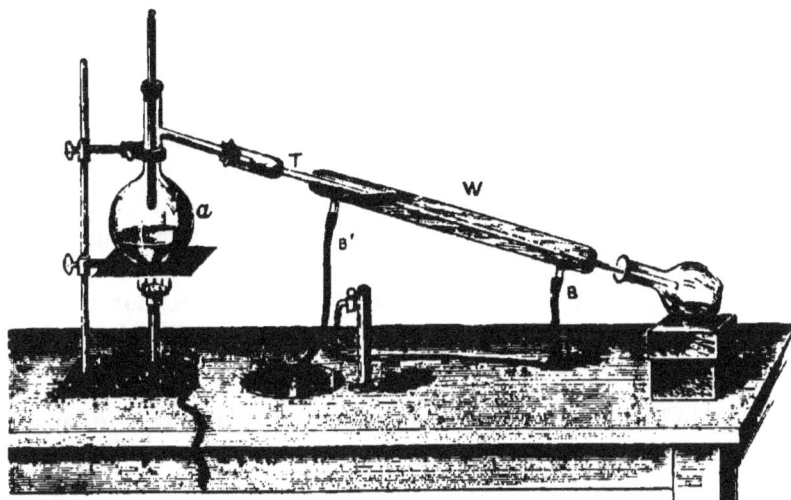

Notice that the mercury gradually rises until a point is reached when it remains stationary. Note this temperature. Observe also that the distillate is perfectly colourless, all the impurity which rendered the water in the flask dirty remains behind, and only clean water passes over. The process of distillation therefore purifies the water; and *distilled water* is purer than ordinary water.

Distillation, also, often enables chemists to *identify* a liquid. For instance, a colourless liquid is given you, which looks like water; if it be submitted to distillation, and the temperature noted at which the mercury in the thermometer is stationary while the liquid is briskly boiling, you could tell at once whether it was water or not.

The process of distillation is useful also as a method of separating liquids which are mixed, but which boil at different temperatures.

Experiment 30.—Place in the Wurtz flask a mixture of about two parts of water and one of alcohol, and have ready three small clean flasks to receive the distillate in. When this mixture is distilled, observe that the mercury in the thermometer quickly rises to a certain point and then remains steady for a time. Presently it begins again to rise ; then change the receiver and collect what passes over separately. At length the mercury is again steady, indicating the temperature at which the water in Exp. 29 distilled ; at this point exchange the second receiver for the third, and continue the process a little longer.

In receiver No. 1 the liquid consists mainly of alcohol, but mixed with a little water. Pour out a little of the liquid and set fire to it, thus proving that it is fairly strong alcohol. Receiver No. 2 contains a mixture of alcohol and a large proportion of water. This liquid will not burn at all. The third receiver contains water free from alcohol.

By this process, therefore, we have obtained a portion of *one* of the liquids, namely, the water, free from the other liquid, but have not completely separated each from the other

Collecting Gases.—Gases are usually collected by causing them to bubble through water into a bottle or cylinder filled with water, and standing mouth downward in a trough or basin of water. This method is usually described as *collecting over water*, and the basin used is called a *pneumatic trough*.

FIG. 9.

Experiment 31.—Fill a glass cylinder to the brim with water, and slide on to the ground lip a ground glass plate so that no air bubble is included. Grasp the cylinder with the left hand, as shown in Fig. 9, and hold the glass cover in its position with the forefinger of the right hand. Then invert the cylinder, when it will be in the position shown in Fig. 10, and lower its mouth beneath the surface of the water in the basin or trough, as in Fig. 11, and now withdraw the plate. The cylinder then remains filled with water. In the figure a *glass* basin is shown, this is in order to render the position of the cylinder visible ; in the laboratory it is more usual to use pneumatic troughs made of metal, having a movable metal shelf with an oblong hole in it through which the bent end of a tube can be introduced. Take a piece of glass tube, bent as shown in Fig. 12, and gently blow through it so that the air from the lungs shall bubble up into the cylinder. As the gas ascends in bubbles, it gradually *displaces the water,* until at last the cylinder is full of gas. To remove the cylinder in order to examine the gas, slip the ground glass plate beneath its mouth, and. keep it in its place with the finger while the cylinder is being turned over. Now remove the plate, and lower a lighted candle on a wire, or a burning taper bent as shown in Fig. 13, into the gas in the cylinder. Notice that the flame is extinguished, showing that the gas which

FIG. 10.

FIG. 11.

comes out of the lungs is different from the air which is drawn into them, because it will not allow a candle to burn in it.[1]

[1] The nature of this gas will be described later on.

Some gases cannot be collected over water, because either they *dissolve* in water, or they enter into *chemical combination* with water. In such cases the *mercurial pneumatic trough* may be used : that is to say, the liquid metal mercury is used instead of water, provided the gas is without chemical action upon mercury.

When a gas happens to be either much lighter or much heavier than air, we may collect it without using a pneumatic trough at all.

Experiment 32.—Obtain a ground glass plate with a hole in the centre,[1] and fit the hole with a cork carrying two bent tubes, one

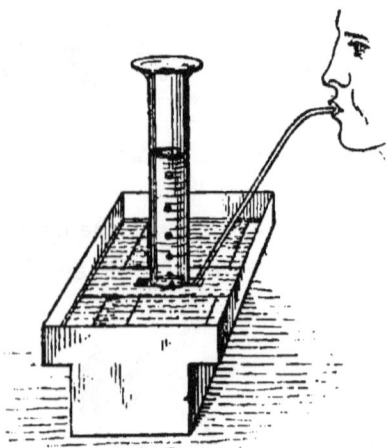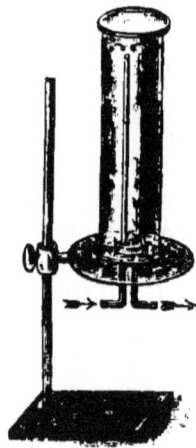

| FIG. 12. | FIG. 13. | FIG. 14. |

long, and the other quite short, as shown in Fig. 14. This is supported on a ring, and the cylinder to be filled is stood over the tubes. Attach the *long* tube to the ordinary coal-gas supply, and turn on the gas. Coal-gas is very much lighter than air (the fact that it is used for filling balloons shows this), and therefore, by delivering it right up to the top of the cylinder, it accumulates in the upper part of the vessel, and gradually displaces the whole of the air, driving it down through the other tube. We can tell

[1] A stout piece of cardboard will answer the purpose, although the glass plate is preferable.

when the air is all displaced by smelling the gas which will then be escaping at the exit tube.

This method of collecting is called *upward displacement.*

Gases that are heavier than air can be collected by *downward displacement.* This is exactly the reverse of upward displacement, and is carried out by simply inverting the apparatus in Fig. 14, so that the cylinder stands mouth upwards.

CHAPTER IV.

FITTING UP APPARATUS.

Much of the apparatus required for such experiments as the elementary student will make, he can himself easily put together by means of glass and rubber tubes, corks, and glass bottles.   Figs. 15, 16, and 17 show three typical forms of apparatus used for the preparation of various gases, and every student should fit up these for himself.

Glass tubes.—Glass tubing is manufactured in many

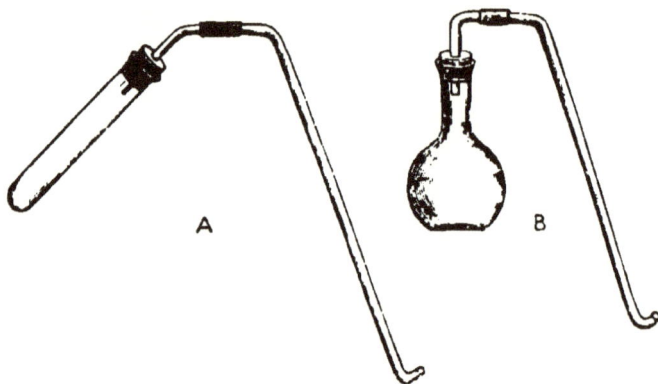

FIG. 15.

varieties of glass, and of many sizes.   For general purposes it is best to obtain *soft soda glass ;* and a most useful size is shown in section at *a*, in Fig. 18.

Such tube is cut to any required length by making a slight scratch upon it with the edge of a fine three-cornered file, and then breaking it across exactly as one would snap a dry twig of wood.

In order to bend a tube, it is made soft by being heated in a common flat gas-flame.

Experiment 33.—Hold a short piece of tube in the flame in the manner shown in Fig. 19, and slowly rotate it between the thumbs and forefingers in order that it may get equally heated all round.

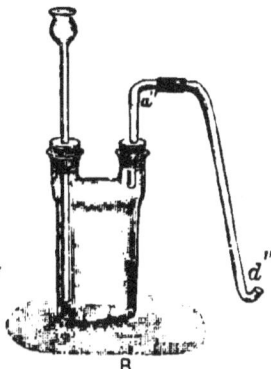

FIG. 16.   FIG. 17.

As soon as it is quite soft, *withdraw it from the flame,* and *deliberately* bend it to a right angle (Fig. 20). When *cold,* wipe off the soot. (Such a right angle bend will be required at a, a', a'' Figs. 16 and 17.)

Tubes should not be heated for bending in a Bunsen flame, as the glass in this case will become *creased* at the bend, as in Fig. 21, B.   If the tubing employed is too thin in the walls (*b*, Fig. 18), the glass will *collapse* at the bend, as shown in Fig. 21, A.   Both of these bends,

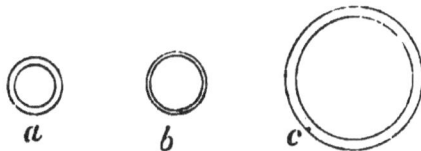

FIG. 18.

besides being unsightly, are very easily broken, and they also prevent the free passage of gas through them.

When a tube is to be bent into a very acute angle, it should be held in the flame in the manner shown in Fig. 22, with

FIG 19

FIG. 20.

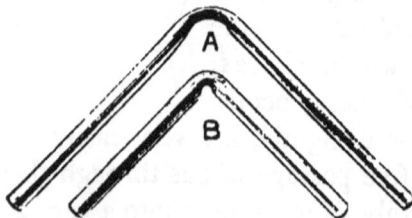

FIG 21.

the knuckles *downwards* instead of upwards, as in Fig. 19, so
that the bend may be produced by bringing the hands together

FIG 22.

FIG. 23.

upwards, as in Fig. 23. Such a bend is required in making
the delivery tubes *d*, *d'*, *d'''*, Figs. 16 and 17. The bent tube

is cut at *c*, Fig. 24, and the end E is then bent as shown by
the dotted lines, by holding it in the flame as explained in
Exp. 33.   All glass tubes, when bent and cut to shape
and length, must be rounded or smoothed at their extremities
by holding them in a flame (Fig. 25) until the "raw" or sharp
edges have just fused.

   "Combustion" tubing is made of hard glass, which will
stand a high temperature without softening.   It is useful when
we require to pass a gas
over some heated material.
c, Fig. 18, shows a convenient
size for combustion tube.

   Corks.—Select a sound
cork which is *just* too large
at its narrow end to fit into
the mouth of the flask or

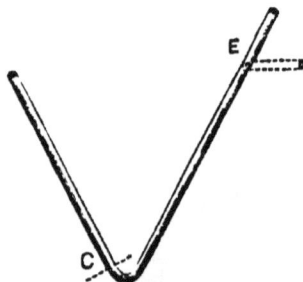

FIG. 24.   FIG. 25.

test-tube, and then squeeze it either in a cork-squeezer or by
rolling it beneath the foot on the floor, using a moderate
pressure upon it.   The cork should then fit comfortably into
the flask.   The cork has next to be bored to take the glass
tube which is to pass through it.   This is best done by means
of a cork-borer, a tool consisting of a brass tube, with a
sharpened edge at one end; it is bored into the cork
much in the same way as a gimlet or brad-awl is used to
bore a hole in wood, except that the cork-borer is usually
wetted.

   In order to be sure that the borer selected will make a
hole the exact size required for the glass tube. a trial hole

should be bored in a waste scrap of cork. The hole must be of such a size that a little gentle force is required to push the tube through it. When using the cork-borer, the tool should be driven in at the narrow end of the cork, care being taken to make the hole in the centre, and to keep it straight. If the borer is fairly sharp, the hole will be clean and smooth.

Sometimes corks are bored by means of a round (or *rat-tail*) file; but this is not a good plan, as it is much more difficult to make the hole perfectly round; and if it is *not* round, the glass tube will not fit properly, and the apparatus will leak at this point.

If the bent tube has been rounded at its ends (p. 32), it can be pushed into the cork without cutting or tearing the hole, and a tight fit will be made.

The delivery tube is attached to the exit tube by means of a short piece of indiarubber tube. The ends of both glass tubes being smoothed, as already described, they can easily be pushed into the caoutchouc connection without cutting it, especially if the latter be moistened inside by breathing through it immediately before introducing the glass.

The apparatus, type No. 2 (Fig. 16), differs only in being fitted with a cork bored with two holes, into the second of which there is fitted a tube known as a *thistle funnel*. This reaches nearly to the bottom of the flask, so that it may dip into the liquids which will be present; and thus, while allowing liquids to be poured in, will prevent gas from escaping through the funnel.

The pieces of apparatus of the first and second types are intended to be used in certain experiments when heat is applied, hence they have to be supported by suitable clamps or stands at a convenient height, and therefore the delivery tubes must be made of such a length that they will reach into the pneumatic trough.

The apparatus of type No. 3 (Fig. 17) is used in the preparation of certain gases, when it is not necessary to apply any heat to the materials. The generating vessel in this case, instead of being a thin glass flask, may be either a two-necked bottle, B, Fig. 17, known as a " Woulf's bottle," or an ordinary

wide-mouthed bottle.   In the latter case, a single cork with two
holes is fitted with two tubes as in A, Fig. 17 ; but when the
two-necked bottle is used the thistle funnel is fitted into one
neck, and the exit tube to the other.   This form of apparatus
is preferable to the other, as small corks are more easily fitted
so as to be free from leakage than wide ones.

When the apparatus has been put together, and before
being used, it should be tested to ascertain whether it is tight.
In the case of A and B, Fig. 15, this is done by sucking a
little of the air out of the apparatus by applying the mouth to
the end of the delivery tube, and instantly closing the tube
with the tip of the tongue.   If the joints are tight the tongue
will remain drawn to the end of the tube, and a little effort is
felt in pulling it away; whereas if there is any leakage in the
apparatus the tongue parts at once away from the tube.

To test the other forms of apparatus in Figs. 16 and 17,
a quantity of water should first be poured into the flask or
bottle, until the end of the thistle tube dips into the liquid.
Then by applying the mouth to the end of the delivery tube
and blowing gently into the apparatus, the water will be forced
up the funnel tube.   It should in this way be driven nearly up
to the head of the thistle tube, and the end of the delivery
tube closed with the tongue.   If the apparatus does not leak,
the water will remain steady in the thistle tube, otherwise it
will gradually sink down.   If the apparatus leaks it should be
refitted, and *on no account should a leaky apparatus be made
tight by the use of sealing wax or other lutes.*

CHAPTER V

GLASS-BLOWING operations, as a general rule, require skill, practice, and patience to perform with anything like success; nevertheless there are a number of smaller and simpler operations which can easily be done by the young student, and which it is most useful that he should know how to perform. The bending of glass tube and the rounding of the ends have already been described. For these operations, however, an ordinary gas-flame is employed, but for those now to be described the blow-pipe is to be used.

A blow-pipe and some sort of foot-blower usually form a part of the regular fittings of a chemical laboratory, but even in their absence much may be done by means of a small Herapath mouth blow-pipe.

General rules.—(1) Never bring a piece of cold glass directly into the blow-pipe flame, but first warm it in the smoky flame before admitting wind from the blower.

(2) When a tube is being heated, it should (except in special cases) be continuously revolved in the flame, so that the heating may be uniform; and also, as it gets soft, to prevent the glass from falling out of shape.

(3) When actually *blowing* glass, always remove the soft glass from the flame.

(4) Always *begin* the blowing gently, and then regulate the force of the breath as the soft glass gives to the pressure.

To open out the end of a glass tube.—When a cork is to be fitted into the end of a glass tube, the end should be opened out a little, or "bordered." The simple tool required for this is a round stick of charcoal, pointed at one end like a lead pencil.

Experiment 34.—Take a piece of moderately wide glass tube and warm one end in the smoky flame (Rule 1). Then admit wind into the flame, and hold the tube in the position shown in Fig. 26, revolving it all the time (Rule 2).

When the end of the glass is sufficiently soft, remove it from the flame, and push the pointed piece of charcoal into it, giving a screwing motion to the charcoal (Fig. 27). When the tube is sufficiently opened, hold it in the smoky flame again, gradually turning down the gas. This *anneals* the glass, that is, cools it slowly, and makes it less liable to crack afterwards.

FIG. 26.

FIG. 27.

To draw down a glass tube to a jet.

Experiment 35.—Heat a piece of tube in the blowpipe flame,

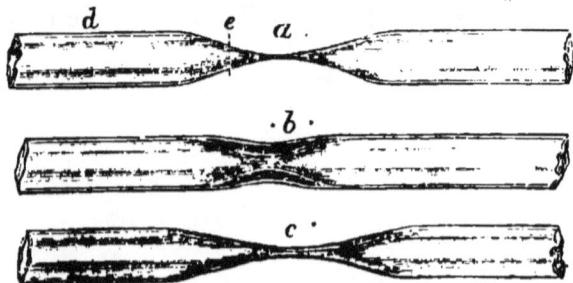

FIG. 28.

holding the glass as in Fig. 19. As soon as it is soft, remove it from the flame, and pull gently, revolving each end slowly at the same time. The glass then assumes the form shown at *a*, Fig. 28, the walls of the tapering and narrow parts being very thin. Now heat another piece, keeping it longer in the flame ; observe that the glass gradually thickens and the walls fall together as the mass gets softer and softer (*b*, Fig. 28). Keep the tube quickly revolving, or the soft part will drop. Very gently draw the ends apart, still revolving, and the tube will take the shape seen at *c*, Fig. 28, where the tapering and narrow parts are thick walled. The tube may then be cut at any desired point on the narrow part by means of a file scratch.

To seal up the end of a glass tube.

Experiment 36.—Draw out a piece of glass tube, as in *a*, Fig. 28. Cut it off at the dotted line *e*, and heat the narrowed end of *d* in the blow-pipe flame until it closes up, when it presents the appearance shown at *f*, Fig. 29. Then heat the somewhat thickened end in the blow-pipe, and, when just soft, blow gently into the tube (note Rules 3 and 4), slowly revolving the glass at the time. It should then appear as seen in *g*, Fig. 29. If too much

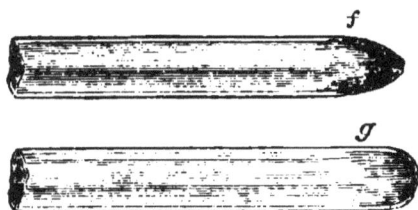

FIG. 29.

pressure was used in blowing, the end will be expanded into a swelling. It may, however, be reduced by revolving it in the blow-pipe flame, when the sides of the enlarged part will fall together again.

To join two tubes.

Experiment 37.—Take two pieces of tube of the same diameter, close one end of one with a cork, and heat the opposite end in the blow-pipe. At the same time heat one end of the other tube. When the ends are soft, bring the two together with a little pressure. This causes them to adhere,

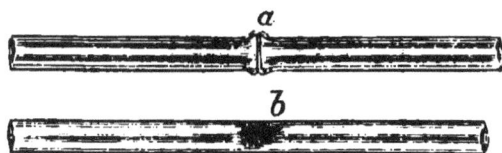

FIG. 30.

and, at the same time, slightly to bulge out at the junction, as at *a*, Fig. 30. Then, while still soft, blow gently into the tube, drawing

it out slightly at the same time so as to keep the outer walls
parallel.   If the blowing operation has not followed the first sufficiently quickly, the joint, as it is at *a*, may be re-heated in a fine-
pointed blow-pipe flame, and then blown as described.   The joint
should have the appearance shown at *b*, Fig. 30.

Experiment 38.—Join a wide tube to a narrow one.   First draw
out the wider tube, and cut it off when it has a diameter equal to
the narrow tube.   Then heat the drawn-out end of the wide tube,
and one end of the narrow tube, and join them in the way described in Exp. 37, blowing through the narrower tube.

To seal platinium wire into glass tubes.   (*a*) In the
end of narrow tubes.

Experiment 39.—Draw the tube out to a point, and cut it off
so that the wire can just pass through the drawn-out end.   Then

FIG. 31.

introduce the tip into a flame, when
the edges of the glass will close together round the wire, as in Fig. 31.
While the glass is still soft, the
position of the wire can be adjusted,
so as to get it quite straight.

(*b*) In either the end, or the side of a wide tube.

Experiment 40.—Heat the glass, at the point where the wire is
to be introduced, with a fine-pointed flame, and when a small spot
is soft, stick the end of a platinum wire into it and draw out gently.
In this way a tiny branch tube is made, *a*, Fig. 32.   This is then
cut off short, as at *b*.   The wire is then inserted, and the flame
again directed upon it, when, as in Exp. 39, the glass closes up
round the wire, *c*, Fig. 32.

To blow a bulb on the end of a tube.

Experiment 41.—First seal up the tube as shown at *f*, Fig. 29.
Then heat the extreme end where the glass is thick, and gently
blow it out so as to obtain the result seen at *a*, Fig. 33.   Next hold
the tube in a large blow-pipe flame, heating mostly the part between
the dotted lines ; and, as the glass softens, keep quickly revolving it.
It then assumes somewhat the shape seen at *b*, Fig. 33.   Withdraw
it from the flame, and blow steadily into the tube, holding it in a
horizontal position, and revolving the glass all the time, *c*, Fig. 33.

[Probably the first attempts will either be complete failures, or very remarkably shaped bulbs, but with a little patience, better results will soon follow.] To blow a larger bulb, more glass is necessary,

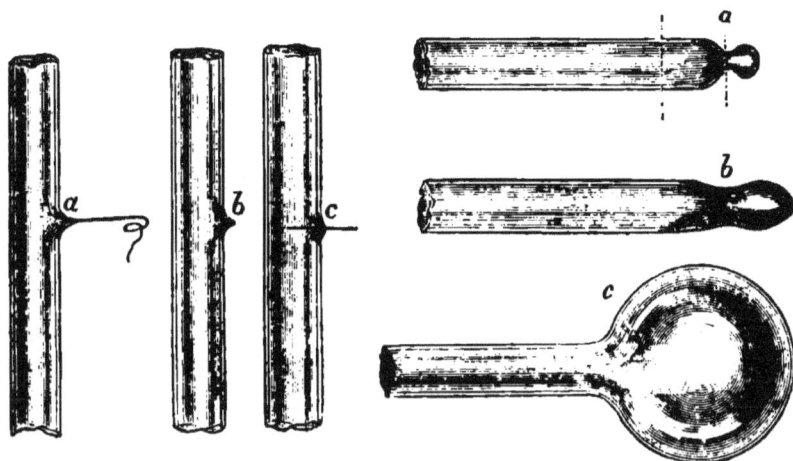

FIG. 32.

FIG. 33.

and it is better to first join on a piece of larger tube as described in Exp. 38.

To blow a bulb on the middle of a tube.

Experiment 42.—Close one end of the tube with a cork, and heat in a large flame at the spot where the bulb is to be blown. As the glass softens and thickens, gently press it together so as to ac-

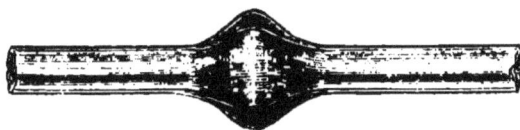

FIG. 34.

cumulate material for the bulb, Fig. 34. Then blow steadily, holding the tube in a horizontal position, and revolve it between the fingers.

CHAPTER VI.

HYDROGEN.

ALTHOUGH solids and liquids are more familiar to us than gases, it will nevertheless be more convenient to begin the study of chemical facts by considering some of the methods of preparing, and a few of the more prominent properties of the gaseous element hydrogen. Let us understand at the very outset, that chemists cannot *create*; they cannot make hydrogen from either *nothing*, or from materials which have got no hydrogen in them. The alchemists of old believed in the transmutation of the metals, and they spent their lives endeavouring to change the common metals into gold; nowadays we know how futile the attempts were, and we no more expect to convert copper into gold, than we expect to "gather figs from thistles."

All that the chemist can do is to so experiment upon compounds which contain hydrogen as one of their constituents, as to cause chemical changes to take place which will result in the expulsion of this hydrogen. We must, therefore, select some suitable compounds containing hydrogen from which to obtain this element.

Chemists have found out, by numberless experiments, that hydrogen forms one of the constituents of a vast number of compounds; thus it is found to be present in nearly all animal and vegetable substances. It is a constituent of water, and also of all those things chemists term acids. There are three common compounds from which the element is most usually obtained; these are—(1) Water, (2) Sulphuric Acid, (3) Hydrochloric Acid.

(1) **Hydrogen from Water.**—As water is the commonest of these three substances, we will first experiment with it. Water is composed of the two elements, hydrogen and oxygen, chemically united together, and in order to separate them by chemical processes, we must find some element which, under suitable conditions, can overcome the force uniting the oxygen and hydrogen, some element which can seize the oxygen and tear it away, so to say, from the grasp of the hydrogen. Many metals are capable of doing this.

Experiment 43.—Carefully throw a small fragment of sodium, about the size of a pea, upon some cold water in an ordinary dinner-plate. Notice that the metal at once melts, and the globule swims to and fro upon the surface of the water (much in the same manner that a drop of water runs about upon a hot iron), producing a hissing sound. Observe that the globule quickly gets less and less, and finally disappears. Now take a strip of turmeric paper (that is, blotting-paper which has been dyed yellow with turmeric) and dip it first into clean water ; this causes no stain upon it ; now dip it into the water in the plate, and observe that it is strongly stained brown. Also dip the fingers into the water, and notice that it feels slimy, or *caustic*, to the touch. This shows that there is something in the water after the sodium has been in contact with it which was not previously there. One of the products, therefore, of the action of the metal sodium upon water is something which dissolves in the excess of water present, yielding a solution which turns yellow turmeric brown, and is caustic to the touch. The other product of the action was hydrogen gas, which escaped unnoticed into the air. If we repeat the experiment, using potassium instead of sodium, the hydrogen will not escape our observation.

Experiment 44.—Throw a similar fragment of potassium upon some clean water in a plate, and at once cover the whole with a glass bell jar, as in Fig. 35. The potassium appears to take fire the moment it touches the water, but really it is the *hydrogen* which

FIG. 35.

burns,[1] the hydrogen which is driven out of its combination with oxygen. The action of the potassium upon water develops so

---

[1] A little of the potassium burns also, and this gives to the flame of the burning hydrogen the violet colour.

much heat as to set fire to the hydrogen. Carefully notice that for a few seconds after the flame goes out, a little red-hot globule of something in a melted state remains swimming upon the water, and then suddenly disappears with a little splutter. (*It is in order to prevent this substance from being scattered, and injuring the eyes, that the plate must be covered with the bell glass.*) Test the water in the plate with turmeric paper, and the same stain is produced as in the former experiment. Therefore, when potassium acts upon water, hydrogen gas is expelled, and a substance is formed which also stains turmeric.

We must now adopt some device for *collecting* the hydrogen gas which can thus be expelled from its combination with oxygen, by the action of sodium upon the water. There are various ways of doing this, but only one of them is free from danger ; for sodium, when brought carelessly into contact with water, is liable to give rise to a serious explosion.

Experiment 45.—Take a piece of lead pipe, $2\frac{1}{2}$ centimetres long (1 inch) and $\frac{1}{2}$ centimetre bore, and close up one end by hammering the lead, as seen in Fig. 36. This little pipe is then filled with sodium by first rolling a pellet of the metal (which is about as soft as wax) between the fingers until it will just push into the tube, and then forcing it in by pressing the mouth of the tube firmly down upon the table. The excess of sodium round the mouth is then trimmed off with a knife. When this tube, with its little charge of sodium, is dropped into water in a dish or trough, it, of course, sinks to the bottom, and a stream of gas-bubbles will be seen rising through the water ; and if the mouth of an inverted glass cylinder, filled with water, be brought over the ascending bubbles, as in Fig. 37, the gas will collect in the upper part. If the sodium is all used up before the cylinder is full, a second leaden tube can be placed in the water. Now remove the cylinder, as described (page 25), and apply a lighted taper to the mouth of the jar. Notice that the gas burns quickly, with rather a yellowish flame, this yellow colour being due to a trace of sodium.

FIG. 36.

If the water in the trough be tested with turmeric paper, it will stain it brown, as in the former case. This other product of the action of sodium on water, which remains dissolved in the large excess of water in the dish, is called sodium hydroxide, or caustic soda; while that which was formed in the case of potassium was potassium hydroxide.

Many other metals besides sodium and potassium are able to expel the hydrogen from water if the *conditions* are favourable. Thus—

Experiment 46.—Drop a few fragments of magnesium into cold water in a test tube, observe that no action takes place between

FIG. 37.

the metal and the water; in this respect, therefore, magnesium is a contrast to sodium or potassium. Now boil the water, and notice that even at the boiling temperature practically no hydrogen is given off.

This experiment shows that magnesium is not able to decompose water appreciably even at the boiling point. Let us, therefore, try the effect of heating the metal much more strongly in a current of steam ; that is, of heating it in contact with water in the state of gas.

Experiment 47.—Fold up a short strip of magnesium ribbon, and place it in a hard glass bulb which is held in a slightly inclined position by a clamp, as in Fig. 38. Attach to one end of the bulb tube a small empty flask, in the manner shown in the figure, and connect this to a tin can in which water is being boiled. As soon as steam issues from the bulb tube, gradually heat the latter all along, by means of a Bunsen flame, until the inside is quite dry, and the steam no longer condenses as it passes through. Then hold the flame steadily under the magnesium, and heat it until it is nearly red hot, when it will suddenly take fire and burn in the steam, combining with the oxygen of the steam and letting the hydrogen go free. Directly this takes place, light the hydrogen as it escapes from the end of the tube. Notice that it burns with a scarcely visible flame, because magnesium does not impart

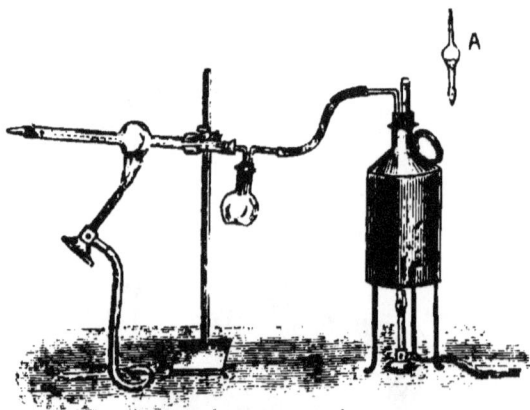

FIG. 38.

any colour to a hydrogen flame like sodium and potassium do. Notice also that the residue in the bulb is white.

Shake out some of this and see if it will dissolve in water by adding water to a little in a test-tube. Note that it is not perceptibly soluble. Take a little more of it and place it on a piece of turmeric paper, and moisten it with a drop of water; notice that the paper is stained slightly brown, showing a slight resemblance between the behaviour of this substance (magnesium oxide or magnesia) and sodium hydroxide.

The common metal iron, at a bright-red heat, will also expel the hydrogen from water when the latter is in the condition of water vapour or steam. This method of obtaining hydrogen is often made use of on a large scale.

Experiment 48.—Take a piece of ordinary iron gas-pipe, sufficiently long to project 15 cm. (6 inches) beyond each end of the furnace to be used. Fill the pipe with small iron nails, and fit each end with a cork with a short straight glass tube. To one end a delivery tube is attached, while the other is connected to a small steam boiler (conveniently made out of a common tin can, see Fig. 38). The iron pipe is heated to a bright-red heat, either in a gas or coke furnace, and steam is passed through it. The red hot iron seizes the oxygen of the steam, combining with it to form iron oxide, which remains in the tube, while a rapid stream of hydrogen is evolved, and this may be collected over water in the pneumatic trough.

(2) **Hydrogen from Sulphuric Acid.**—Although water is the commonest compound of hydrogen, sulphuric acid is the one from which it is most convenient to obtain hydrogen, and when we require this gas for experiments, it is almost always got from sulphuric acid by the action upon it of the metal zinc.

Experiment 49.—Place some granulated zinc [1] in a two-necked

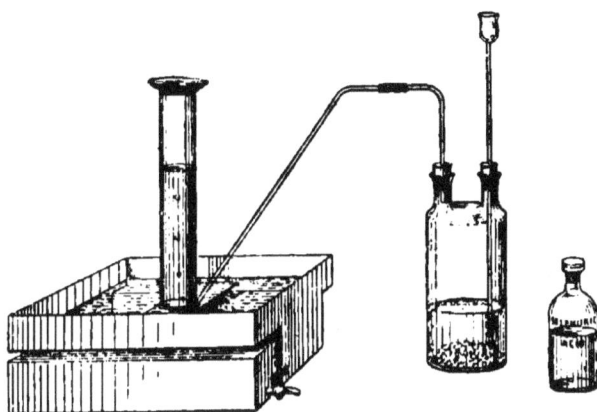

FIG. 39.

Woulf's bottle, arranged as in Fig. 39, with just sufficient water to cover it, and pour upon it by means of the thistle funnel a little strong sulphuric acid. Almost immediately it will be noticed that an effervescence begins, showing that gas is being disengaged. Allow the action to go on for a few minutes until all the air originally present in the bottle has been swept out by the gas, and then collect the gas at the pneumatic trough. Three or four cylinders should be filled for subsequent experiments.

In this experiment the metal zinc expels the hydrogen from sulphuric acid, and combines with what is left of the

[1] Granulated zinc is made by first melting the metal and pouring it in a thin stream into cold water.

sulphuric acid after the hydrogen is gone. The compound so produced, called zinc sulphate (or *white vitriol*), is left dissolved in the water in the bottle. It may be obtained from the solution by the following experiment.

Experiment 50.—When the above experiment is concluded, pour the liquid out of the Woulf's bottle and filter it. Then gently evaporate it in a dish over a small flame until it is reduced to about half the bulk, and allow it to cool, when long glassy-like colourless crystals will deposit. This is the zinc sulphate.

Instead of employing zinc in Exp. 49, the metal iron might have been used; but the hydrogen obtained in this way is always contaminated with compounds of hydrogen with both carbon and sulphur (both of these elements always being present in ordinary iron), which give to the hydrogen an unpleasant smell.

Experiment 51.—Put a small quantity of iron filings into a test-tube, and pour upon them a little dilute sulphuric acid. Notice effervescence, due to the escape of gas. Smell the gas, and note the peculiar and nasty odour. This smell does not belong to *hydrogen*, but to the impurities present. Bring a lighted taper to the mouth of the tube; observe that the gas burns. When the action has continued for some time, the liquid may be filtered, and the clear solution evaporated, when small green crystals will be deposited. This substance is the compound of iron with what is left of the sulphuric acid after the hydrogen in it is expelled. It is called ferrous sulphate (or *green vitriol*).

In a similar manner we can expel the hydrogen from sulphuric acid by means of the metal magnesium.

Experiment 52.—Drop a few fragments of magnesium ribbon into a little dilute sulphuric acid in a test-tube. Notice how briskly the gas is given off. Show that it is hydrogen by lighting it at the mouth of the tube. When the metal is all dissolved, this solution also may be evaporated down and allowed to cool, when colourless crystals of magnesium sulphate will be formed. Compare these crystals with those of zinc sulphate.

(3) **Hydrogen from Hydrochloric Acid.**—The hydrogen from this compound is also expelled by the metals, zinc, iron, and magnesium; therefore, hydrochloric acid may be

used instead of sulphuric acid in Exps. 49 to 52. The compounds which remain behind in the solution would, in this case, be compounds of the metals with what is left of hydrochloric acid after the hydrogen has been expelled; they would be zinc chloride, ferrous chloride, and magnesium chloride respectively.

The Properties of Hydrogen.—From the various samples collected, we learn that the gas is colourless (therefore invisible). Also that it does not appreciably dissolve in water, for when left standing in vessels in the pneumatic trough the water does not show any signs of rising up in the cylinders, which it would if the gas were soluble. If we take one of the jars of gas and smell it, we shall find that it is quite odourless. We have also learnt from Exps. 45 and 47 that hydrogen will burn.

It is important to remember that if we take *any* gas which will burn in the air, and previously mix it with a certain proportion of air, and then bring a light to the mixture, an explosion results. Every one knows that it is dangerous to bring a light into a room where there is an escape of coal-gas, that is, where there is a mixture of gas and air. Coal-gas *alone* burns quietly; but a *mixture of coal-gas and air* in certain proportions explodes when lighted. This is particularly true of hydrogen. We have seen by Exp. 47 that, when unmixed with air, it burns quietly, but if a mixture of hydrogen and air be lighted, the mixture explodes with great violence. Therefore, before bringing a flame to the tube leading from a hydrogen apparatus, it is *most important* to be sure that all the air originally present in the bottle has been swept out. In order to realize the force of the explosion that would occur under such circumstances, and the danger which would result from neglecting this precaution, the following experiment may be made—

Experiment 53.—Fit an ordinary pear-shaped soda-water bottle with a cork, through which is fixed a short piece of the stem of a clay tobacco-pipe, only just projecting through the cork. Put a little granulated zinc into the bottle, add some dilute sulphuric acid, and insert the cork. Hold a lighted taper to the end of the tube as though attempting to inflame the hydrogen, and in a few seconds, when the gas has mixed with the air in the bottle in a certain proportion, a loud explosion will result, which will shoot the cork out of the bottle with some violence. There is no fear of the thick bottle bursting, but it should be held in such a position that the cork will not fly in a direction where it can do harm.

Experiment 54.—Disconnect the bent delivery tube from the hydrogen apparatus, and by means of a rubber tube attach a straight glass tube. Fill a test-tube with the gas by *upward displacement*, as in Fig. 40. After a few moments withdraw the delivery tube, and apply a light to the mouth of the test-tube. If the hydrogen is mixed with air, a slight *pop* will result, but if free from air it will burn quietly. If this is the case the gas may be lighted at the end of the delivery tube. Notice that the flame has a yellowish colour, this is due to the soda present in the glass (recall the appearance of the flame when the hydrogen collected in Exp. 45 was burnt). The true appearance of a flame of hydrogen may be seen by sub-

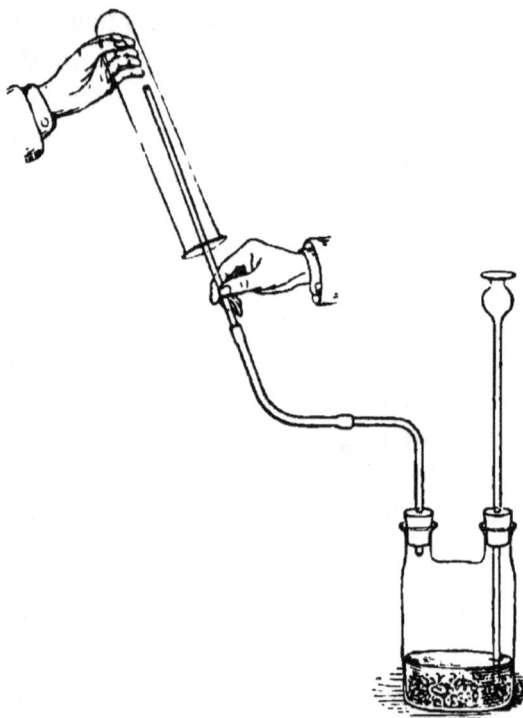

FIG. 40.

stituting for the glass tube a short piece of lead pipe, into one end of which an ordinary metal gas-burner has been screwed. Notice that the flame now is almost without colour, being slightly bluish, and gives no light.

Place an ordinary coal-gas flame by the side of the hydrogen flame and compare them. Depress into each flame a white plate, observe that the bright flame blackens the plate, while no soot is deposited from the other. Hold a clean dry jar over each flame for a few moments, notice that in each case the flames are giving off *steam*, because moisture will be deposited on the cold sides of the glass. Cover each of the jars with a glass plate, or piece of card, and pour into each a little clear lime-water. After shaking the lime-water with the air in the jars, it will be seen that in the jar which was held over the gas flame, the lime-water has turned milky, while in that which was over the hydrogen it remains clear.

These experiments teach us that *water* is formed when we burn either hydrogen or coal-gas ; and that from burning coal-gas we also get, besides the water, a gas which will turn lime-water milky.

Hydrogen is the lightest substance known to chemists. It is nearly $14\frac{1}{2}$ times lighter than air. On account of its extreme lightness it can be collected in vessels by upward displacement (see Fig. 41). It may also be *poured upwards* from one vessel to another.

Experiment 55.—Take a cylinder containing only air, and hold it mouth downwards in the left hand ;

FIG. 41.

then take another cylinder filled with hydrogen in the other hand, and bring its mouth just beneath that of the first, in the manner shown in Fig. 41, and gradually empty its contents up into this one by lowering the foot until the position shown in Fig. 42 is reached. Now stand the lower cylinder down, and apply a light to the contents of first one and then the other. Notice that no hydrogen is left in the one which formerly was full, and that the contents of the upper jar burn with the characteristic flame.

Owing to the extreme lightness of hydrogen, it is sometimes used for filling balloons. On a small scale we can imitate this by filling soap bubbles with the gas.

Experiment 56.—Attach a common clay tobacco-pipe to the hydrogen apparatus, and proceed to blow a bubble, holding the pipe in the usual position. Notice that the bubble in trying to

E

ascend, very soon begins to curl over on to the outside of the pipe, as in Fig. 43. Turn the pipe upside down, and blow another, and observe that as it grows it struggles to tear itself away from the pipe, assuming the shape seen in Fig. 44. When disengaged from the pipe it ascends very quickly.

FIG. 42.

A pretty experiment is to make a small *hydrogen* bubble carry up a large *air* bubble. To do this, first blow an ordinary bubble with the breath. Then, while it is still hanging on the pipe, bring under it a hydrogen bubble just beginning to form (Fig. 45). As the latter increases in size, gradually invert the two pipes so that the one holding the hydrogen bubble is uppermost. Then, with a slight jerk, detach first the pipe from the air bubble, and next the one conveying hydrogen. The double bubble then slowly rises (Fig. 46), and generally when it touches the ceiling the little hydrogen one breaks, and allows the other to fall to the ground again. This experiment requires a good soap solution.[1]

FIG. 43.

Although hydrogen will burn, it will extinguish the flame of any ordinary burning substance. This is true of all gases

---

[1] The following is a good receipt. Dissolve 2 grams of sodium oleate in 80 cc. cold distilled water, add 20 cc. glycerine, well shake, and put away in a dark cupboard for two days to settle. Then carefully pour off the clear liquid, add just 1 drop of ammonia and shake up.

FIG. 44.

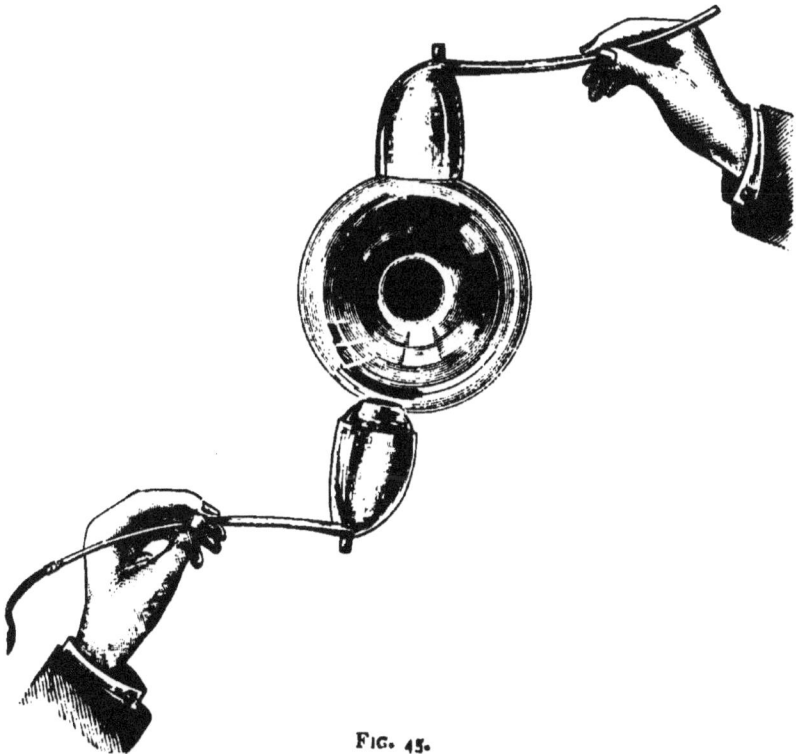

FIG. 45.

which burn in the air, that they will put out the flames of all other things which burn in the air.

Experiment 57.—Thrust a lighted candle (or piece of thick taper), fastened to the end of a wire, into a cylinder of hydrogen held mouth downward. As the burning candle approaches the mouth of the vessel, it there sets fire to the hydrogen ; but as it is pushed up into the gas its *own* flame is extinguished. Withdraw the candle and relight it as it passes out through the still burning hydrogen.

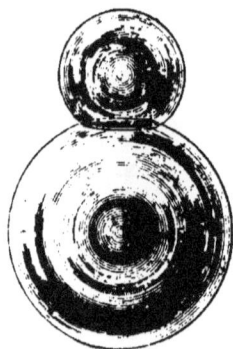

The same will happen if a paper spill be used instead of a taper, or if an ordinary coal-gas flame be introduced.

Fig. 46.

We therefore say that hydrogen is a *combustible* gas, but that it is a *non-supporter of combustion.*

EPITOME.

The element hydrogen is found *free* in only very small quantities on the earth ; but it is present on the sun in enormous quantities. It occurs in nature in combination with oxygen in water ; with carbon in marsh gas ; with sulphur in sulphuretted hydrogen. It is present in nearly all animal and vegetable substances, and is a constituent of all *acids.*

It can be obtained from water : (1) by the action of either sodium or potassium at the ordinary temperature ; (2) by the action of magnesium or iron at a red heat.

It may be got from either sulphuric acid or from hydrochloric acid by the action of either zinc, iron, or magnesium upon them.

The common laboratory method is by acting on sulphuric acid with zinc.

Hydrogen is a colourless, tasteless, odourless gas. It burns with a nearly invisible bluish flame.

A mixture of hydrogen and air explodes when lighted. It extinguishes the flames of other ordinary burning substances.

Hydrogen is the lightest of all known substances, being 14·4 times lighter than air. For this reason it is taken as the standard or unit for comparing the densities of all other gases. Thus we say that the density of air is 14·4, which means that air is 14·4 times heavier or denser than hydrogen, bulk for bulk. Hydrogen being thus taken as the unit, obviously its density is unity or 1.

Reactions for hydrogen—[1]

(1) From water, by the action of sodium $H_2O + Na = NaHO + H$.
(2)   ,,   ,,   ,,   magnesium $H_2O + Mg = MgO + H_2$.
(3)   ,,   .,   ,,   iron $4H_2O + 3Fe = Fe_3O_4 + 4H_2$.
(4) From sulphuric acid, by the action of zinc $H_2SO_4 + Zn = ZnSO_4 + H_2$.
(5)   ,,   ,,   ,,   magnesium $H_2SO_4 + Mg = MgSO_4 + H_2$.
(6)   ,,   ,,   ,,   iron $H_2SO_4 + Fe = FeSO_4 + H_2$.
(7) From hydrochloric acid, by the action of zinc $2 HCl + Zn = ZnCl_2 + H_2$.
(8) The combustion of hydrogen in air or oxygen $H_2 + O = H_2O$.

---

[1] These signs and symbols will be explained later on; the student may pass them over at this stage.

CHAPTER VII

THIS element, like hydrogen, is a gas, but in almost every other respect it presents a complete contrast to that element. There are a great many compounds containing oxygen as one of their constituents, from which the element can easily be obtained, for in combination with others it is the most abundant of all the elements. Unlike hydrogen, it is found in the free or uncombined state in large quantities on the earth, for the atmosphere consists essentially of free oxygen, *mixed* with about four times its volume of nitrogen.

We shall consider the methods of obtaining this element from four of its compounds, namely, from—

(1) Mercuric oxide ; (2) Potassium Chlorate ; (3) Sodium peroxide ; (4) Water ; and also the way by which it is obtained from the air.

(1) **Oxygen from Mercuric Oxide.**—We saw in Exp. 6 that when this compound is simply heated, it is decomposed into its two constituent elements, oxygen and mercury. In order to collect the gas which is given off we proceed as follows—

Experiment 58.—Heat a small quantity of the red powder in a hard glass tube, arranged as shown in Fig. 15, A, and collect the gas over water. Notice that the evolution of gas is not very rapid ; also that, as in Exp. 6, the metal mercury collects on the cooler part of the tube in small globules.

The chief interest in this experiment lies in the fact that

it was the very method by which Priestley first discovered oxygen in 1774. He called it *dephlogistigated air ;* the name oxygen was given to it later by Lavoisier.

(2) **Oxygen from Potassium Chlorate.**—When this salt is heated, it first melts and then rapidly gives off oxygen. It is decomposed by heat into two things, namely, into oxygen, which passes off as gas, and into potassium chloride, which remains behind as a white solid.

Experiment 59.—Heat a small quantity of potassium chlorate in a similar apparatus to that used for the last experiment. Notice

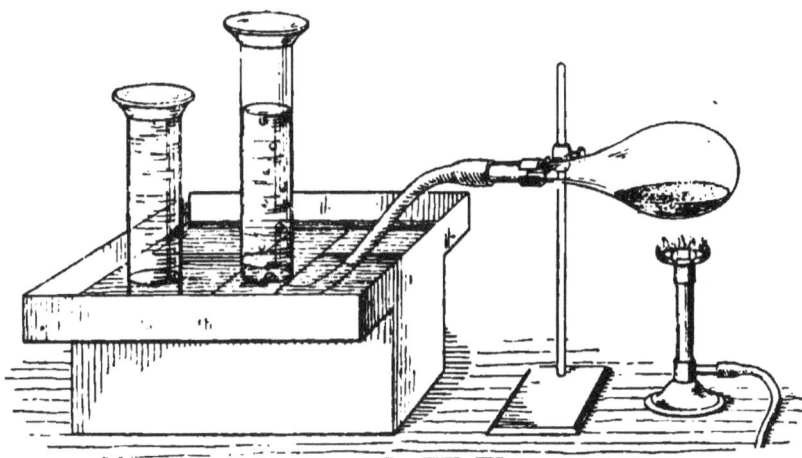

FIG. 47.

that the crystals first crackle, then melt, and that the melted compound then begins to effervesce, owing to the escape of the oxygen gas. Note also that the gas is given off much more rapidly than in the case of the mercuric oxide.

It has been found that if the potassium chlorate be previously mixed with manganese dioxide, the chlorate gives up its oxygen much more rapidly, and at a much lower temperature.

Experiment 60.—Mix about 20 grams of potassium chlorate with about a quarter of its weight of powdered manganese dioxide, and gently heat the mixture in a flask (a common Florence oil flask) as shown in Fig. 47. Notice first that the mixture does not melt ; also that moisture collects in the neck of the flask. This chiefly comes from the manganese dioxide, which is always damp ; and it is in order to prevent this condensed moisture from running back into the heated flask and cracking it, that the apparatus is supported in a horizontal position.   Observe how rapidly the gas is evolved, and with what a little heat.   It is to allow plenty of passage for the gas that the usual glass delivery tube is here replaced by a wide piece of indiarubber tube.

Notice that during the experiment little sparkles occasionally appear in the heated mixture.   These are caused by small particles of combustible impurities which are always liable to be present in manganese dioxide.[1]

Collect several cylinders or jars with the gas as it is evolved, and keep them standing mouth downward in small plates containing a little water.

This is the method usually employed in the laboratory for preparing oxygen.

The material left in the flask after the experiment consists of potassium chloride and manganese dioxide.   The latter substance therefore comes out of the reaction in exactly the same state as it was at the beginning.   If the mixture is boiled with water, the potassium chloride dissolves and leaves the manganese dioxide ; so that if the mixture is then filtered, the dioxide can be recovered, and used over and over any number of times.   The *way* in which it acts in causing the potassium chlorate to give up its oxygen more readily, involves a series of rather complex changes, which cannot be conveniently considered at this stage.

(3) **Oxygen from Sodium Peroxide.**—When this compound is brought into contact with water it is at once decomposed, and oxygen is evolved.

Experiment 61.—Place a small quantity of sodium peroxide in a dry flask fitted with a delivery tube and a stoppered funnel, as in Fig. 48.   Allow water to enter the flask drop by drop by means of the funnel.   As each drop falls upon the powder, a brisk action is

---

[1] If a *large* quantity of such impurity were present, such as would occur if the black oxide of manganese were adulterated with powdered coal, it would give rise to an explosion when heated.

noticed, and oxygen is rapidly given off, which may be collected in the usual way.

After the experiment, dip a piece of turmeric paper into the liquid in the flask ; notice that the paper is stained. Also dip the fingers into it, and note its caustic nature. Recall the solution obtained by the action of sodium upon water (see Exp. 43). Here we have the same substance, caustic soda, formed as a second product of the chemical change.

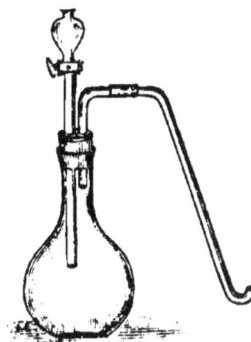

FIG 48.

(4) Oxygen from Water.—Although water is the commonest compound of oxygen, we very seldom employ it for obtaining this gas, because water is not so easily decomposed as many other oxygen compounds. There are very few elements which have a sufficiently strong affinity for the hydrogen of the water, for them to tear it away from the oxygen, and let the latter go free. Under certain conditions, however, the element chlorine (also a gas) is able to do this. Thus, if a mixture of chlorine gas and steam be strongly heated by being passed through a red hot tube, the chlorine seizes the hydrogen of the water, unites with it to produce the compound hydrogen chloride (or hydrochloric acid), and the oxygen in the water is set free.

(5) Oxygen from the Air.—Oxygen is now obtained on a manufacturing scale from the enormous store of it which is present in the air. This is done in two operations ; in the first the atmospheric oxygen is made to combine with some substance, and in the second this substance is again decomposed. The substance employed is barium oxide (*baryta*). This is heated in iron pipes through which air is pumped under a slightly increased pressure. Under these circumstances the barium oxide combines with oxygen from the air, and gives barium dioxide, while the other chief constituent of the air, namely, the nitrogen, passes away. Presently the pumps are reversed, and a partial vacuum is produced in the heated pipes ; this causes the barium dioxide to decompose, changing back into the original barium oxide, and giving up the oxygen it had absorbed from the air. Therefore, by pumping air in and out of these heated pipes containing baryta, it is possible to obtain large quantities of oxygen very rapidly. This method of obtaining oxygen is known as "Brin's process."

The Properties of Oxygen.—From the various speci-mens of the gas which have been prepared, we see that it is colourless, and that it is so little soluble in water that we observe no loss while collecting it in the pneumatic trough. If we take one of the jars and apply the nose to the gas, we shall find that oxygen has no smell or taste. This is exactly what we might expect, when we remember that the air we breathe contains a large proportion (one-fifth) of free oxygen.

Test for Oxygen.—Oxygen is generally recognized and distinguished from other gases by thrusting into a jar of it a chip or splinter of wood which has been lighted, and has only a glowing spark upon it. The splinter is instantly rekindled and bursts into flame. There is, however, one other gas (namely, nitrous oxide) which behaves in a similar manner towards a glowing splint. Therefore this test does not dis-tinguish oxygen from nitrous oxide. How these two gases are identified is explained on p. 202.

Experiment 62.—Test the samples of gas obtained by heating mercuric oxide, and from sodium peroxide, by plunging into them a splinter of wood which has a glowing spark upon the end. The splinter will almost immediately rekindle. Blow it out again, and, while the end is still glowing, thrust it once more into the gas. It again bursts into flame as before. This can be repeated so long as sufficient oxygen is left in the jar.

All substances which are capable of burning in the air, will burn more rapidly and with increased brilliancy in pure oxygen. It is entirely on account of the oxygen present in the atmo-sphere, that substances burn in the air at all, and it will be evident that they must burn more quickly in oxygen *alone*, than in oxygen which is diluted with a large quantity of nitrogen, as it is in the atmosphere. We may prove this by burning a number of substances in the gas already prepared.

Experiment 63.—**Charcoal in Oxygen.** Take a piece of charcoal about the size of a hazel-nut, and fasten it, by means of a piece of thin copper-wire, to a deflagrating spoon. Light one corner of the charcoal in a gas-flame, and then lower the spoon into a jar of oxygen. Notice that the charcoal at once begins to burn much more brightly than it did before. If the charcoal used, happens to be a piece made from the *bark* of the wood, it will throw off a shower of brilliant sparks or scintillations as it burns in the oxygen. When the charcoal has burnt out, remove the spoon, and pour a little water into the jar. Cover it with a glass plate and keep it for a further experiment.

Experiment 64.—**Sulphur in Oxygen.** Unscrew the small cup from a deflagrating spoon, and tie a small bundle of asbestos to the end of the wire by means of thin copper wire. Then melt a little sulphur in a test-tube and dip the asbestos into it so as to get it coated over with the sulphur. Now light the sulphur upon the asbestos by means of a lamp flame, allow it to burn in the air for a moment or two and plunge it into a jar of oxygen. Observe the greatly increased brilliancy of the flame the moment it comes into the oxygen. Notice that the burning sulphur produces a little smoke or fume in the jar. When the sulphur has burnt out, pour a little water into the jar, shake it up, cover the jar over with a glass plate, and keep it for a subsequent experiment.

Experiment 65.—**Phosphorus in Oxygen.** Take a piece of phosphorus [1] about the size of a pea, wipe it dry with blotting paper, and place it in a deflagrating spoon. Set fire to the phosphorus by touching it with the end of a wire which has been just warmed in a flame, and lower the spoon into a large flask filled with oxygen. Observe how much more intense is the light of the phosphorus burning in oxygen than burning in the air. Watch the experiment closely, and notice that, as the burning goes on, suddenly the light becomes almost unbearably dazzling, and then quickly dies down. At that point the phosphorus was rapidly boiling.

FIG. 49.

In preparing to do this experiment, the wire of the deflagrating spoon should be pushed so far through the metal cap that the cup containing the phosphorus reaches *nearly*

---

[1] Remember the precautions as to handling phosphorus given on p. 10.

to the bottom of the flask as shown in Fig. 49, so as to prevent the flame from cracking the flask at its shoulder.

When the phosphorus is burnt out, remove the spoon, add a little water, and shake it up in the flask. Notice that the dense white fumes in the flask rapidly disappear ; they are dissolved by the water.

Experiment 66.—Pour a little blue litmus solution into the jars used in Exps. 63, 64, 65. Notice that the blue solution is turned red in each case ; but observe that in the jar in which the carbon was burnt the red colour is less of a scarlet, and more inclined to purple than in the other two cases.

Substances which have the power of turning litmus red, are said to be *acid;* therefore, when we burn carbon, sulphur, and phosphorus in oxygen, and add water to the products of the burning, we obtain substances which are *acids.*

Experiment 67.—Sodium in Oxygen. Put a small fragment of sodium in a clean dry deflagrating spoon, and heat it in a gas flame until the sodium begins to burn in the air, then plunge it into a cylinder of oxygen. The sodium burns very brightly, and produces a white smoke. The product of the combustion is, however, in this instance a solid, and most of it, therefore, remains behind in the spoon. When the sodium has all burnt, remove the spoon, and rinse out the jar with a little water, and pour the liquid out into two small beakers. To one add a little litmus, and notice that it is *not* turned red. To the other add a little litmus which has been just turned red by a single drop of dilute acid, and observe that this reddened litmus is turned back to its original blue colour.[1]

Substances which restore the blue colour to reddened litmus are said to be *alkaline;* therefore, when sodium is burnt in oxygen and the product dissolved in water, we obtain an *alkali.*

Many substances which will not burn in the air, will burn readily in pure oxygen.

Experiment 68.—Iron in Oxygen. In order to burn iron in oxygen, it must, like the phosphorus or the sulphur or the charcoal, be first lighted. To do the experiment it is best to use a jet of

[1] If the liquid obtained by rinsing out the jar does not contain enough of the product from the burning sodium to show this result, the main portion which was left in the spoon may be used, by dissolving it off in a little water.

oxygen direct from a store of the gas contained either in a gasholder or better in a metal cylinder, into which the oxygen has been pumped under great pressure.

Connect to a gas reservoir by means of a rubber tube, a glass tube, drawn to a jet at one end, and allow a stream of oxygen to blow a spirit lamp flame against the ends of a bundle of fine steel wires, in the manner shown in Fig. 50. Almost immediately the tips of the wires get red hot, and begin to burn. Now remove the lamp and the metal will continue to burn in the jet of oxygen, throwing off a shower of brilliant sparks. Notice that the end of the bundle of wires melts, and drops of molten matter continually fall. These should be allowed to drop into an iron basin of water placed ready to catch them.

In the absence of a reservoir of oxygen, fill a common wide-mouthed bottle (such as a pickle bottle or glass marmalade jar) with

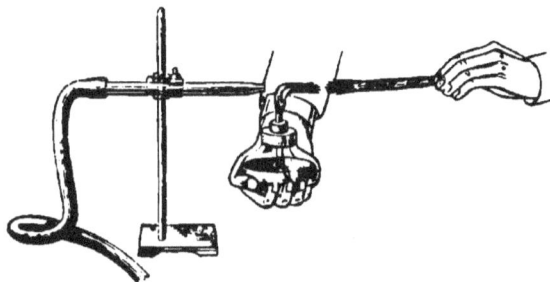

FIG. 50.

oxygen, leaving about two inches of water in the bottle. Stick into a loosely fitting cork a bundle of steel wire or a straightened piece of watchspring, the end of which has been tipped with sulphur. Light the sulphur in a flame and quickly plunge the wire into the bottle of oxygen. The burning sulphur will set fire to the iron, which will go on burning in the oxygen. The melted product of the combustion will be partially quenched by the layer of water ; but, in spite of this, it will probably crack the bottle.

Experiment 69.—Collect some of the solid product of the burning iron, and place it upon a moistened litmus paper. Notice that the paper is not turned red.

Redden another piece of litmus paper by dipping it into water containing a drop or two of acid, and place some of the substance upon this. The blue colour is *not* restored ; therefore the product obtained by burning iron in oxygen is neither acid nor alkaline.

In each case the product obtained when a substance is

burnt in oxygen, is a compound of the thing burnt with the oxygen. The process of burning in these instances is therefore nothing more than the *rapid* combination of the various substances with oxygen. If we were to burn carbon, sulphur, phosphorus, or sodium in the air, and examine the products obtained, we should find that in all cases they were the same as when these things were burnt in oxygen. This shows that the ordinary process of combustion in the air is also the rapid combination of substances with oxygen, the difference being that the combination is not so rapid as in pure oxygen. We can easily prove that it is the oxygen present in the air which enables substances to burn in air, by removing the oxygen from a quantity of air, and trying the experiment of putting a lighted candle into the air that is left.

Experiment 70.—Place a fragment of phosphorus (wiped dry) in a little dish, and float the dish, upon water in a pneumatic

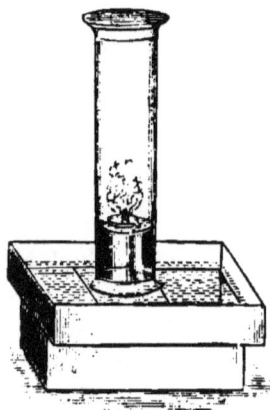

FIG. 51.

trough. Set fire to the phosphorus, and cover it with a cylinder a little wider than the small dish as in Fig. 51. The phosphorus burns in the confined air in the cylinder, and combines with the oxygen present, forming the same white fume as when burnt in pure oxygen. When the phosphorus has burnt itself out, allow the apparatus to stand for a short time, so that the jar may cool, and the fumes may become dissolved by the water. Notice that the water has risen in the cylinder, taking the place of the oxygen that has been withdrawn. Now slip a glass plate beneath the mouth of the cylinder and remove it from the trough. Shake the air and water together so as to completely dissolve the remaining fume. Now introduce a lighted taper or candle into the gas in the jar, and notice that the flame is at once extinguished. A lighted spill of paper, or a coal-gas flame, will behave in a similar manner. We have withdrawn all the oxygen from the air in the cylinder, and the gas that remains (namely, nitrogen) will not support the combustion of ordinary burning bodies.

Oxides.—The products obtained when substances are

burnt in oxygen (either pure oxygen, or the oxygen of the air) are called *oxides*. All the elements, except fluorine, have been made to combine with oxygen either directly or indirectly.

We have seen from Exps. 66, 67, and 68, that different oxides have very different properties ; thus, some combine with water to yield acids,[1] while others, under the same circumstances, give compounds which are alkaline. The former of these are called *acid-forming oxides*, while the latter belong to a class known as *basic oxides*. (As a general rule the non-metals combine with oxygen to give acid-forming oxides, while the metals yield basic oxides. There are, however, some oxides of both non-metals and of metals which are neither acid-forming nor basic.)

Hydroxides.—This is the name applied to the compounds which are produced when oxides combine with water. The acid-forming oxides yield *acid hydroxides*, while those derived from basic oxides are called *basic hydroxides*. The term *acid hydroxide*, however, is not very often employed, as these compounds are included in the class of substances called *acids ;* and the name *hydroxide* alone is more usually used to denote the basic hydroxides.

Since the basic hydroxides are derived from oxides of *metals*, they consist of a metal combined with oxygen and hydrogen : they are the hydroxides of metals. Thus, sodium combined with oxygen and hydrogen (or sodium oxide combined with water) is sodium hydroxide ; calcium united with oxygen and hydrogen (or calcium oxide united with water) is calcium hydroxide ; and so on.

Oxidation.—The process of combining with oxygen is called *oxidation*, whether the action takes place rapidly, as when substances burn in oxygen, or whether it goes on slowly without any visible signs of heat. We saw in Exp. 61, the *rapid* oxidation of the metal sodium, but the same process of oxidation goes on if sodium is merely exposed to oxygen, even the oxygen of the air, without being made to *burn*.

---

[1] For further development of this subject, see p. 66.

Experiment 71.—Take a fair sized piece of sodium and quickly cut a slice off it. Notice that for an instant the freshly cut surface of the metal looks bright and silvery, but that almost immediately it becomes dull and *tarnished.* It quickly gets coated with a white film or crust which consists of the oxide of sodium. The tarnishing is the *oxidation* of the metal, and if the piece of sodium be left exposed to the air it is soon oxidized right through, the whole piece being changed into the oxide.

Other metals behave in a similar way; thus, when bright iron is exposed to the air, we know that it soon loses its brilliant surface, and becomes coated with a reddish film of *rust.* This rust is simply the *oxide of iron,* formed by the slow combination of the metal with the oxygen of the air.

Increase of Weight by Burning.—If the ordinary process of burning is simply the rapid combination of the burning substance with the oxygen of the air, we ought to find that the products of burning actually weigh more than the material that is burnt. This we can easily put to the test of experiment.

Experiment 72.—Place a heap of "reduced iron"[1] on a little iron dish or tray, and balance it upon a small pair of scales. Then, by means of a lighted taper, set the heap alight. Observe that the black powder gradually smoulders right through, and turns to the familiar brownish colour of iron rust; but note also that as it burns the mass gains in weight, for the scale pan on which it is soon begins to fall. This shows that the rust is heavier than the iron which produced it.

In this experiment the only product of the burning is solid and visible, but when we burn a candle we see no products. The candle simply appears to waste away, leaving nothing behind; by burning it we seem to have completely annihilated it. Is this really the case? or can it be that the products of the burning of a candle are *gases,* and escape unobserved into the air? Let us try the following experiment.

---

[1] Reduced iron is readily prepared by heating oxide of iron in a glass tube, and passing a stream of coal-gas or hydrogen through the tube until the material is black.

Experiment 73.—Take a piece of wide glass tube (such as a lamp chimney) and fill the upper part with lumps of solid caustic soda, which are kept in position by first hanging a false bottom of wire gauze into the tube by means of wires. Fit a cork into the bottom, on which is fastened a short candle, the cork being bored with holes to allow air to enter (Fig. 52). Balance this arrangement upon a pair of scales, and then light the candle and quickly replace the cork. Again note that as the candle burns the apparatus becomes heavier. The candle in the act of burning is combining with oxygen, and two invisible gaseous products are formed, namely, steam and carbon dioxide, both of which are caught or *absorbed* by the lumps of caustic soda.

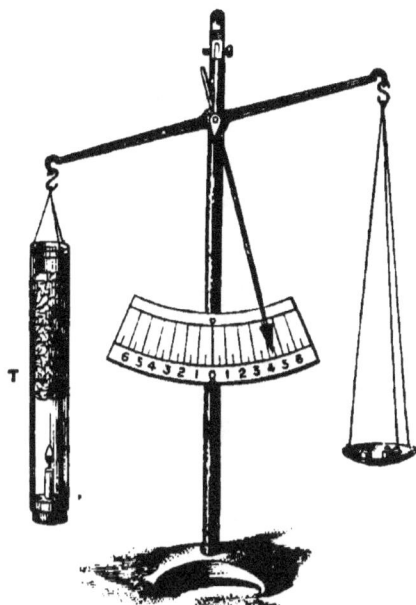

FIG. 52.

Matter is Indestructible.—We can destroy a *candle* by burning it, but not the *carbon* and *hydrogen* of which the candle is composed; these live on, as it were, but in different states of combination. Similarly if we burn a piece of sulphur it disappears from view, but the sulphur is not *destroyed*, only passed into combination with oxygen, forming a compound which is gaseous and invisible.

Respiration.—Not only is oxygen necessary to support the combustion of ordinary burning bodies, but it is also indispensable to the process of respiration. If an animal is placed in air from which the oxygen has been removed, or in any mixture of gases which does not contain free oxygen, it quickly dies. All animals require oxygen to breathe. Respiration is, in fact, a process of oxidation; air is drawn into the lungs, and a portion of the oxygen is absorbed by the blood. This oxygen-laden blood (which has a bright red colour) passes throughout the body, and exerts its oxidizing power

F

upon certain compounds containing carbon which have to be
removed from the system, with the result that carbon dioxide
is produced (the same compound as is formed when carbon
is burnt in oxygen, or when a candle burns in the air) and is
exhaled in the breath. When the red blood has thus parted
with its oxygen it becomes of a dark colour, and is known
as *venous* blood; this, travelling back to the lungs, is there
once more charged with oxygen, again to carry it to all parts
of the body. We can, by a simple experiment, show that the
air exhaled from the lungs contains the same gas as is formed
when a gas flame or a candle burns in the air.

Experiment 74.—Place some clear lime-water in a test-tube,
and by means of a glass tube dipping into it, bubble the breath
from the lungs through it. Note that the *first* portions of breath
produce little or no effect, but that presently, as air from the deeper
parts of the lungs is expelled, the clear solution quickly becomes
milky.

And we can also show that the air so expelled has been
deprived of some of its oxygen by the following experiment.

Experiment 75.—Collect in a cylinder over water the breath
exhaled by one single long expiration, emptying the lungs as far
as possible (see Exp. 31, Fig. 12). Now remove the cylinder and
lower into it a lighted candle on a wire. Notice that the flame will
be extinguished. [This result is due partly to the presence of the
carbon dioxide, as well as to the smaller amount of oxygen.]

Acids.—There is a class of compounds which chemists
call *acids*. All acids have a sour taste, and they also have the
power of reddening a solution of blue litmus (refer back to
Exp. 66). It does not follow, however, that all sour substances,
or all that will redden litmus, are *acids*. This is by no means
the case, for we know of many things which have a sour taste
and which will turn litmus from blue to red, but which do not
belong to the class of substances recognised as *acids;* they are
acid, in the sense of being sour, but are not *acids*.

The question, What is an acid? is one which cannot be
answered in a word, and chemists are not agreed as to what
is the best definition to be given. At one time it was supposed

that all acids contained oxygen, that this element was the *acidifying principle* in these substances. Indeed the very name *oxygen* means the *acid producer*, and was originally applied to this element by Lavoisier because of this idea. We now know that this belief was wrong, for we have many acids which do not contain any oxygen in their composition. At the present time chemists regard the element *hydrogen* as a necessary constituent of an acid. According to modern notions, a compound which does not contain any hydrogen cannot be an acid. This of course does not mean that all compounds containing hydrogen are acids. We know, for example, that water is a compound of hydrogen, but water is not an acid, it is not sour, and does not redden litmus. We have already learnt that sulphuric acid and hydrochloric acid contain hydrogen, because we obtained this element from both of them by acting upon them with various metals. Can we, therefore, define an acid as a compound from which hydrogen can be expelled by the action of certain metals? No, because we have also learnt that hydrogen can be expelled from *water* by means of certain metals, and water is not an acid. Perhaps the best definition that can be given is the following, *an acid is a compound from which hydrogen can be displaced by the hydroxide of a metal.* The word *displaced* here only means that the hydrogen is *turned out of the compound*, and not that it is set free, or liberated as gas.

Alkalies.—This name is applied to the hydroxides of the alkali metals (of which sodium and potassium are the most important) and to ammonia. When sodium is burnt in air or in oxygen, and the oxide so obtained is dissolved in water, sodium hydroxide is produced (Exp. 67). We have learnt that the alkali produced in this experiment has the property of changing the colour of turmeric from yellow to brown, and also of restoring the blue colour to litmus which has been reddened by an acid.

Experiment 76.—Make a dilute solution of sodium hydroxide (caustic soda) by dissolving a small piece of the solid in water in a beaker. Dip a turmeric paper into it, and observe the stain. Make a dilute sample of hydrochloric acid by adding a little strong

acid to some water in a beaker. Dip a turmeric paper into this ;
note that no stain is produced. Add to the acid a few drops of a
solution of litmus, which will be at once changed from blue to red.
Now add the dilute alkali very gradually to the acid, and watch
the effect upon the reddened litmus. Where the alkali meets the
acid solution, the litmus shows a blue colour, which, however, disappears again on gently shaking or stirring. Presently, however,
as more alkali is added, the red liquid turns entirely blue. The
solution is now no longer *acid,* but *alkaline.* Now add, drop by
drop, a little more of the dilute acid, stirring the solution ; stop the
moment the liquid turns red. If this is carefully done, one drop
of the alkali will now restore the blue colour. Now test this
solution with a turmeric paper, and note that no brown stain is
produced ; therefore the liquid is *neither acid nor alkaline.*

Neutralization.—Exp. 76 teaches us that when we mix
an acid and an alkali, each one destroys the other, so
that a point can be arrived at when the solution has neither
the property of an acid nor an alkali. Under these circumstances we say that the liquid is *neutral,* that the acid has
neutralized the alkali, or the alkali has neutralized the acid.
Now we must ask ourselves what has become of the acid and
the alkali? Are both still present, and the properties of each
simply masked by those of the other? In other words, have
we simply a *mixture* of these things so adjusted that the
properties of the one just balance those of the other, or has
any chemical change taken place between the acid and the
alkali resulting in new compounds which happen to have no
effect upon either turmeric or litmus? Let us test this matter
by experiment.

Experiment 77.—Take some of the dilute acid, the alkali,
and the neutral mixture, and evaporate each separately to dryness
in small dishes. Note first that the acid leaves no residue behind :
that is, it is a volatile acid ; therefore, if the neutral solution is
simply a *mixture* of the two substances, when it is evaporated
down, the acid ought to volatilize along with the steam and leave
the alkali, in which case the residue in the dish containing the
alkali, and that in the one containing the neutral solution would
be the same. Are they the same? Observe that they *appear*
different. Touch each with a moist turmeric paper, the one is

strongly alkaline, while the other is still perfectly neutral, therefore they *are* different. The residue being neutral shows that by evaporating the solution, the acid, which had been added, has not been driven off; it cannot, therefore, have been simply mixed with the alkali, but must have entered into chemical union with it. Now, with the little finger bring a little of the neutral residue on to the tip of the tongue; the familiar taste of the substance will also prove that we have here a compound which is quite different from either sodium hydroxide or hydrochloric acid.

We learn, therefore, that when acids and alkalies neutralize one another they enter into chemical combination with each other, forming new compounds.

Bases.—A large number of other substances besides the alkalies, are capable of neutralizing acids. Many of these substances resemble the alkalies in having the power of restoring the blue colour to reddened litmus, and in imparting a brown colour to turmeric; they are, therefore, said to be *alkaline* in character, or to possess an *alkaline reaction.* Some chemists, indeed, extend the name alkali so as to include many of these compounds. The term *base* has long been in use to denote any substance which is capable of neutralizing an acid, and it therefore includes the alkalies. For the most part bases are compounds of metals; being either the oxides or hydroxides of metals. Ammonia, however, which contains no metal, being only a compound of nitrogen and hydrogen, is also included in this class, and so are many organic compounds which are not compounds of metals. For our present purpose, however, we may define bases as *certain oxides and hydroxides of metals, which are able to neutralize acids,* and include *ammonia.*

Salts.—The compounds that are produced when acids and bases combine together are called *salts.* We have seen by Exp. 77 that when hydrochloric acid and sodium hydroxide neutralize each other the substance we familiarly call "salt" is produced. "Salt" is one of the commonest and most important of all salts, as well as being one that has been longest known to man; and the compounds belonging to this class are called *salts,* because of a general similarity many of them bear to "salt," or *common salt,* as it is termed.   No satisfactory definition of a salt can be given, because chemists are not all agreed as to exactly what shall be included in this class of compounds.   Some regard acids themselves, as being *salts of hydrogen.*   At this stage we will consider salts as being the *compounds that are produced by the interaction of acids with bases.*

EPITOME.

Oxygen was discovered in 1774 by Priestley, by heating mercuric oxide.   It is the most abundant of all the elements.   Occurs uncombined in the air.   It is usually obtained by heating potassium chlorate: more readily if the chlorate be mixed with manganese dioxide.   Oxygen is also given off when certain peroxides, such as manganese dioxide, or barium dioxide, are strongly heated ; the latter substance is used in Brin's process.   Oxygen can be obtained from water by heating steam and chlorine together.

Oxygen is a colourless, tasteless, and inodorous gas ; supports combustion energetically, and is necessary to life.   Oxygen is slightly soluble in water ; 100 volumes of water can dissolve 4 volumes of oxygen.   Fish depend upon this dissolved oxygen for their supply of this gas for respiration ; they cannot breathe free air, neither can they use the oxygen which is a chemical constituent of water.

The products obtained by burning things in oxygen, or in air, are oxides.   Some of these are acid-forming, others are basic oxides.   Similar compounds are formed when the same elements combine slowly with oxygen, the process then being called oxidation.   The products obtained by burning or by oxidation are heavier than the original substance that is burnt ; burning is only matter undergoing change, and not matter being destroyed.

Acids are compounds containing hydrogen which can be displaced from the compound by the hydroxide of a metal.   They are sour or acid to the taste, and will redden blue litmus.

Bases are compounds which will *neutralize* acids.   They are mostly certain oxides and hydroxides of metals.   Ammonia is also a base.   Those that are soluble in water are alkaline ; they restore the blue colour to reddened litmus, and give a brown stain to turmeric.

Salts are the compounds produced by the union of acids and bases.

Reactions for oxygen [1]—

(1) From mercuric oxide $HgO = Hg + O$

(2) „ potassium chlorate $KClO_3 = KCl + 3O$

(3) „ sodium peroxide and water $Na_2O_2 + H_2O = 2NaHO + O$

(4) „ water by the action of chlorine $H_2O + Cl_2 = 2HCl + O$

(5) „ air, by Brin's process—(*a*) $BaO + O = BaO_2$

(*b*) $BaO_2 = BaO + O$

Combustions in oxygen—

Carbon $C + O_2 = CO_2$

Sulphur $S + O_2 = SO_2$

Phosphorus $2P + 5O = P_2O_5$

Iron $3Fe + 4O = Fe_3O_4$

[1] These will be explained later on; the student can pass them over at this stage.

CHAPTER VIII

WE have learnt by Exp. 54 that when hydrogen burns in the air, water is formed; and as other elements on burning in the air yield their *oxides*, so we might expect that hydrogen would do the same, and that water is an oxide of hydrogen. First let us make an experiment in order to collect the liquid which is formed, and examine it.

Experiment 78.—Burn a small flame of hydrogen under the open end of a bent glass tube, arranged as in Fig. 53. A liquid quickly condenses on the long neck and runs down into the flask, and in about half an hour a considerable quantity will be collected.

One property by which we can recognize water from other liquids we have learnt from Exp. 44, and that is its behaviour towards the metal potassium.

Experiment 79.—Pour the liquid collected in Exp. 78 into a test-tube, and drop into it a small bit of potassium; if the metal takes fire, as it did in Exp. 44, we conclude that the liquid is water. [Other tests by which we can recognize water we shall learn later on.]

As the air does not consist of oxygen alone, we cannot say *certainly* from this experiment that water is a compound of hydrogen and oxygen only. This point might be settled by burning the hydrogen in pure oxygen. It was indeed first settled for us by Cavendish, in 1781.

Cavendish used a strong glass vessel, similar to E, Fig. 54, which had a stopcock at the bottom, and was closed at the top with a stopper through which two wires were fixed, so that he could send an electric spark into the gases it contained. He pumped all the air out of this vessel with an air pump, and then attached it to the bell jar, B, containing a mixture of hydrogen and oxygen. On opening the taps the gases of course entered into the vacuous tube. The stopcocks were then closed, and a spark from an electrical machine was made to pass between the two wires. This of course exploded the

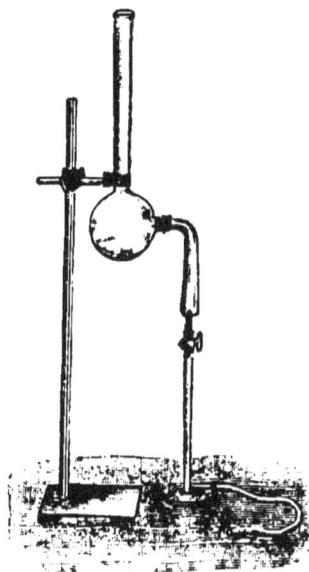

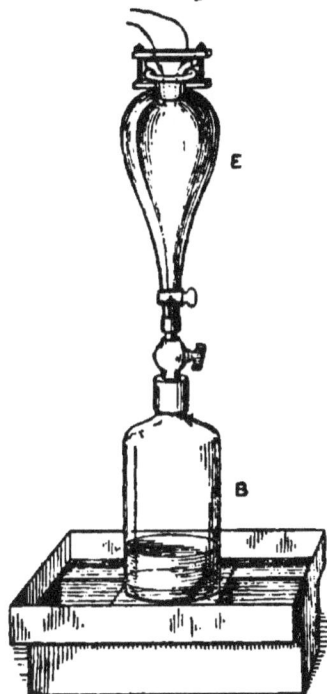

Fig. 53. Fig. 54.

mixture of hydrogen and oxygen, but the vessel being very strong did not burst. *After the two gases had thus combined, there was a minute quantity of water in the tube.*

By again filling the vessel with more of the gas and exploding the mixture, the quantity of water increased, until by repeating the experiment several times, enough of the liquid was collected to prove that it was actually pure water. As there was nothing besides hydrogen and oxygen present in the gases, this proved that water was composed simply of these two elements.

But Cavendish did more than merely prove this, he also

taught us the *proportions* in which these two gases combined
together to produce water.   If, for instance, the mixture of
gases in the bell jar contained equal measures of hydrogen and
oxygen, he found that after the explosion there was always
some oxygen left over in the vessel E.   But if he mixed the
gases exactly in the proportion of two measures of hydrogen to
one of oxygen, then there was no gas left in E after the explosion ; it entirely disappeared, leaving a vacuum in the vessel.
Therefore Cavendish made the discovery that water was a
compound of hydrogen and oxygen only, and that *these two
elements combined in the proportion of two volumes of hydrogen to
one volume of oxygen.*

We can make hydrogen and oxygen combine together by
what are called *indirect* methods, and thus get additional proof
that water is composed of these two elements only.

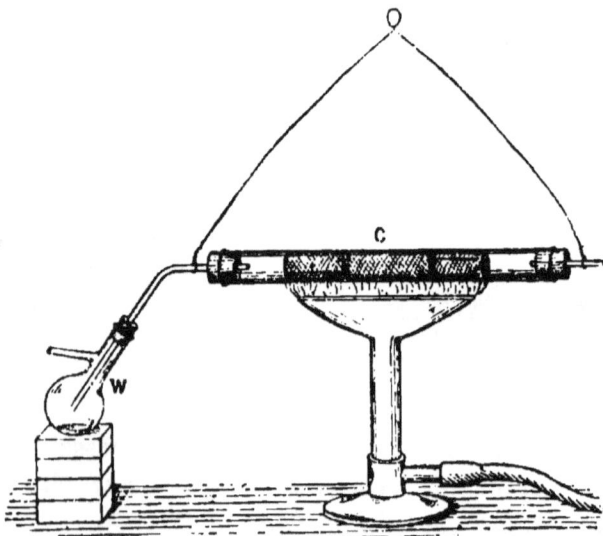

FIG. 55.

Experiment 80.—Roll a piece of fine copper gauze into short
compact cylinders, *c*, Fig. 55, and place them in a piece of hard
glass tube (*combustion tube*).   To one end of this tube connect a
small Wurtz flask, in the manner shown at *w*, Fig. 55, and to
the other attach a small apparatus for generating oxygen from potassium chlorate.   Heat the combustion tube with a long flat flame

until the copper is red hot, and then send a slow stream of oxygen over it by gently heating the potassium chlorate. Notice that as the oxygen passes over the copper, the latter becomes black. It is gradually combining with the oxygen and is being converted into *copper oxide*, which is black. After some little time disconnect the oxygen generator, and replace it by an apparatus for generating hydrogen from zinc and sulphuric acid, and pass a slow stream of hydrogen through the tube.[1] Observe that almost at once some moisture condenses in the Wurtz flask, and as the experiment goes on, more and more water collects in the little receiver. Where is this water coming from? Notice that the black copper oxide is gradually changing back again into bright metallic copper. The hydrogen has taken away the oxygen which the copper had combined with, and has united with it to form water.

In this experiment therefore we have caused hydrogen and oxygen to unite, by first combining the oxygen with copper, and then allowing hydrogen to deprive the copper oxide of the oxygen, whereby the copper oxide was again *reduced* (that is, the copper was deprived of the element with which it had combined, and was restored to its former state of metallic copper), and water was formed. This experiment is a very important one, and we shall return to it again later. In Cavendish's experiment, and in Exp. 78, water was obtained by the *direct* combination of its constituent elements; by Exp. 80, it was formed by the *indirect* union of oxygen and hydrogen; but whether directly or indirectly, the compound was built up, so to speak, from the elements of which it is composed. This method of proceeding is called synthesis.

There is another way by which we can find out the composition of a compound like water, which is exactly the opposite of synthesis. Instead of taking the constituents of the compound and making them combine, we can take the compound and *decompose* it into its constituents. This method is known as analysis.

---

[1] A convenient piece of apparatus, easy to make, by means of which a regulated stream of hydrogen can be obtained at will, is described on p. 246, Fig. 106. Put granulated zinc in one tube and dilute sulphuric acid in the other.

For instance, in Exps. 45, 48, we decomposed water by sodium, and by iron, and got out of the water one of its constituents, namely, the hydrogen.

By the action of chlorine upon water (as described on p. 57), the water was again decomposed, and the other constituent, namely, the oxygen, was obtained. These are processes of *analysis*.

Water can also be decomposed by means of an electric current from a battery.

Experiment 81.—Take an ordinary wide mouthed bottle, and cut it in half. (This may be done by first making a small scratch on the glass with a file, and then touching the spot with a red hot wire.

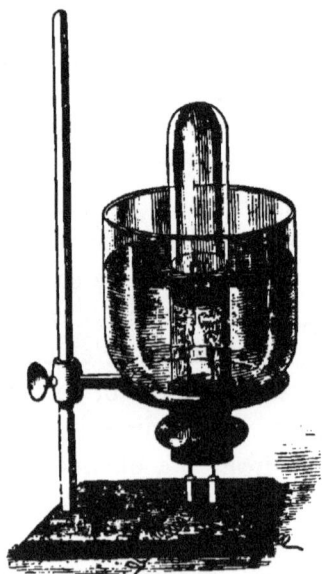

FIG. 56.

The glass will then crack at the file mark, and the crack can be made to travel right round the bottle by slowly drawing the hot wire along just in front of the crack.) Fit into the neck a cork through which two short pieces of platinum wire have been pushed. The apparatus is then supported as shown in Fig. 56, and nearly filled with water, to which a little sulphuric acid has been added.[1] Now attach the two wires from the battery to the two platinum wires, and notice that bubbles of gas at once make their appearance on each of the wires in the acid water. Fill a short stout test-tube with the dilute acid and invert it over the two wires in the basin, and collect the gas that is evolved. When the tube is full, remove it and apply a lighted taper to the gas. If it is hydrogen it will be recognized by the characteristic flame; if oxygen, we shall see the taper flame burn more brightly. Note, however, that the gas does not answer to either of these tests, but that it explodes with a sharp crack. This shows that we have both oxygen and hydrogen mixed together.

[1] Pure water will not conduct electricity, hence sulphuric acid is added to it.

Experiment 82.—Now collect the gas from each wire separately by inverting a separate tube over each wire, using two test-tubes of the same size. Notice that gas collects more quickly in one tube than in the other. As soon as the tube which fills quickest is full down to the water level, stop the experiment by disconnecting one wire from the battery. Note that the other tube is only half full. Now test the gas in each tube, and find that the gas in smallest quantity will rekindle a glowing splint of wood; it is therefore oxygen; while the other burns with the characteristic flame of hydrogen.

These two experiments prove by *analysis* that water consists of oxygen and hydrogen; and also that it contains these two gases in the proportion of two volumes of hydrogen to one volume of oxygen. They therefore confirm Cavendish's *synthetical* experiments.

When oxygen and hydrogen unite (as in Cavendish's experiment), the volume occupied by the water that is formed is so extremely minute, that it is scarcely measurable unless very large volumes of the two gases are used. We cannot, therefore, make any comparison between the volume of the gas and that of the *liquid* water that is produced. But if, instead of letting the water condense to the liquid state, we were to make the experiment at a *high temperature*, so that the water was kept in the condition of *steam*, then we could find out the relation between the volume of the mixed gases, oxygen and hydrogen, and the volume of the gaseous compound, steam.

This experiment is made in the following way. Two measures of hydrogen and one of oxygen are put in the closed limb of the U tube (Fig. 57), which is divided into three equal measures by rings on the glass. The tube is then heated by boiling some amyl alcohol in a flask and sending the hot vapour through the outer glass tube which surrounds the tube containing the gases. The vapour leaves this "jacket tube" by the small pipe at the bottom, and is again condensed and collected. When the whole apparatus is hot, the mercury is made level in both limbs, and the gas exactly occupies the three measures. Now an electric spark from an electrical machine is passed between two platinum wires which are sealed into the closed end of the tube containing the gases This causes the oxygen and hydrogen to combine (which they do with some violence, so that precautions are taken to prevent the mercury from being blown out at the open end of the U tube) and form water, which, however, cannot condense to the liquid state, but remains as steam, on account of the high temperature of the vapour surrounding the tube. The first thing noticed after the explosion of the gases, is that the volume is altered, for the mercury has risen in the tube.

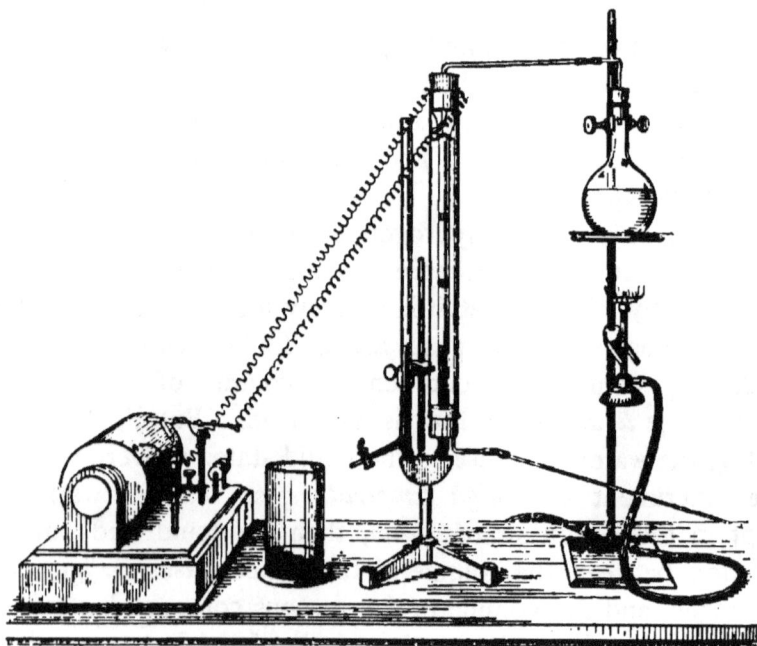

FIG. 57.

Once more, therefore, the mercury must be made level in both limbs, by pouring more into the open side; and when this is done, it is seen that the gas in the tube now occupies exactly two of the measures. That is to say, two volumes of hydrogen and one of oxygen, *three volumes of the mixed gases,* form two volumes of steam or gaseous water.

The Properties of Water.—Water is familiar to us as a colourless and tasteless liquid. When we look through a considerable depth of it, however, it is seen to be possessed of a bluish-green colour. We know that when water is cooled to a certain temperature it solidifies or freezes; and when heated to a particular temperature it boils. One method, therefore, by which we are able to distinguish water from any other colourless liquid is to ascertain its freezing and boiling points. The freezing point of water is 0° on the centigrade scale, while its boiling point is 100° (see p. 88).

As water is gradually cooled down, like other substances it shrinks in volume. This shrinking or contraction goes on steadily until the temperature is just within 4° of the freezing point. After this temperature is passed, the water, instead of continuing to contract, actually expands again, just as though it were being warmed. Therefore, if we take some water having a temperature of 4° C., it will expand whether we heat it or cool it. At this particular temperature, water is more dense than at any other point; 4° C. is, therefore, called its point of maximum density.

The expansion that water suffers on being cooled from 4° to 0° (without freezing) is extremely slight; 1000 cc. of water measured at 4°, when cooled to 0°, will become expanded to 1000·13 cc. But this expansion, although apparently so trifling, plays an important part in nature. When lakes and ponds are exposed to the cold winter winds, the surface water becomes cooled, and consequently contracts and becomes denser. It therefore sinks to the bottom, and fresh layers come to the surface to be cooled in their turn, and also to sink. In this way a circulating movement is set up, which continues until the temperature of the whole mass of water has fallen to 4° C. When this state is reached, any further cooling of the surface *expands* the water on the top. It therefore becomes less dense, and so remains as a colder layer floating upon the surface, until at last it solidifies to a thin film of ice, which then protects the water beneath from contact with the cold wind. If water continued steadily to contract with cold until the freezing point, this circulating movement would not stop at 4° C., but would go on until the whole body of water in lakes and ponds had reached 0°, when ice would begin to form at the bottom as well as the top, and the entire mass of water would very soon become solidified throughout.

When water actually freezes, it suddenly expands very considerably, 10 volumes of water becoming 11 (nearly) volumes of ice. Ice is therefore lighter than water. 10 cc. of water weigh 10 grams, but 11 cc. of ice only weigh 10 grams. Consequently ice floats on water.

The force which is exerted by the expansion when water freezes is very great. Thus, if a strong iron bottle be filled entirely with water, and the mouth securely closed, and then the bottle be exposed to a freezing temperature, the expansion will cause the bottle to burst. It is owing to this that water-pipes burst in frosty weather. The pipe splits, or bursts, at the moment the water freezes, but it is only when the ice melts again that the damage to the pipe is revealed ; hence the mistake is often made of supposing that the *thaw* caused the pipe to burst.

We say that water boils at a temperature of 100° C. ; but, to be exact, we must add *under a pressure of 760 mm.* (see p. 92). The actual temperature at which water, or any other liquid, boils, depends entirely upon the pressure. This is very easy to prove. If we take a quantity of water in a beaker, scarcely hotter than can be borne by the hand, therefore far short of boiling, and place it under the receiver of an air-pump, and reduce the pressure by quickly pumping out some of the air, we shall see the water begins to boil violently, although obviously it cannot be getting any hotter. In a vacuum water can be made to boil even at the freezing temperature, therefore *boiling* has not necessarily anything to do with *being hot.* In ordinary life we are in the habit of associating the two ideas together, so that the very word *boiling* is almost synonymous with being hot; hence the common expression " boiling hot." In reality the expression " boiling cold " is quite as correct, for many boiling liquids are colder than ice, and colder even than the coldest arctic regions.

The Solvent Power of Water.—Water is able to dissolve a great many substances, and chemists make use of this property in many different ways. Thus substances, which in the solid state are incapable of acting chemically upon each other, will often combine if either one or both of them are first dissolved in water, hence a large number of chemical operations (especially analytical) are conducted with solutions in water. The solvent power of water is also made use of for separating substances which are more soluble, from others which are either slightly soluble or are altogether insoluble (see Exps. 20 and 27).

Many gases are soluble in water; some to a very great extent, others only very slightly. Thus, 1 cc. of water at 0° will dissolve 1148 cc. of ammonia gas, but only 0·048 cc. of oxygen. In all cases, gases are *less* soluble in hot than in cold water; thus at 20° 1 cc. of water can only dissolve 680 cc. of ammonia, and 0·028 cc. of oxygen. Therefore, if a solution of a gas in water be boiled, all the gas is expelled.

Natural Waters.—Owing to its great solvent powers, perfectly pure water is never found in nature, and, indeed, can only be obtained in the laboratory by taking great precautions. The purest form of natural water is rain water, but even this dissolves the gases from the air, and is also contaminated with impurities that are present in the air. The moment rain touches the ground, it begins to dissolve the soil or rocks upon which it falls and through which it percolates, and it gradually becomes more and more impure as it finds its way to river and ocean.

Absolutely pure water when evaporated to dryness in a platinum vessel leaves no residue. If boiled down in a glass vessel it slightly dissolves the glass. Sea-water contains more dissolved impurities in the form of various salts than other natural waters. Thus, in 1000 grams (1 litre) of the water of the British Channel, there are 35·25 grams of dissolved salts, which remain as a residue when this volume of the water is evaporated to dryness. The greater part of this residue, viz. 27 grams, is common salt. The solid dissolved substances in good drinking water average from 0·437 to 0·03 grams in 1 litre.

Hardness of Water.—Certain of the dissolved solids in natural waters impart the property known as hardness. These salts are chiefly the carbonate and sulphate of lime. When hard water is boiled, if its hardness is due to the presence of

G

dissolved carbonate of lime, it becomes soft, because the
carbonate of lime is precipitated as a solid, which in time
collects on the sides of the vessel (boiler or kettle), and produces the *furring* which is so common. Sulphate of lime is
not thrown out of solution by boiling ; therefore, if the hardness
is due to this salt, boiling the water will not soften it. On
this account, hardness caused by carbonate of lime is called
temporary hardness ; and that due to sulphate of lime is
termed *permanent hardness.* If a sample of water contains
both these compounds, then it loses only a part of its hardness, viz. the temporary hardness, on boiling, while the
permanent hardness remains.

Water of Crystallization.—When salts are crystallized
from solution in water (as in Exp. 26), it often happens that
some of the water solidifies along with the salt. It is as
though the particles of the salt were unable to build themselves up into the regular shapes, we call crystals, without the
aid of some particles of water to hold them together ; just as
bricks cannot be built up into an edifice without the aid of
mortar. If the mortar were to be removed from a building,
the bricks would all tumble down in a heap ; and if we remove
the solidified water from such a crystal, it will fall down to
a powder. Water which is thus held by a crystal, and which
is necessary to its existence as a crystal, is called water of
crystallization.

Experiment 83.—Take a few crystals of copper sulphate (*blue
vitriol*) and gently heat them in a dish, either over a small flame
or better in an oven. Notice that gradually the crystals lose their
blue colour, and become white, while at the same time they lose
their crystalline nature and are changed to a powder. Heating
the crystals has driven off their water of crystallization, and consequently they no longer retain their shape as crystals.

In the case of the copper sulphate it is easy to see when
the water of crystallization is removed, because the salt itself
(the *anhydrous* salt, as we call it) is nearly white, while the
crystallized salt is deep blue. If a quantity of the white *de-
hydrated* copper sulphate be moistened with water, it is at
once *re-hydrated,* that is, it takes up water of crystallization

again and consequently turns blue. Many other salts change colour when they lose either some or all of their water of crystallization.

Experiment 84.—Dissolve a crystal of cobalt chloride in water. Notice the pink colour of the salt and also of the solution. With a brush, or a clean pen, write on paper, using this solution instead of ink. The writing will be invisible, because the pink colour is so faint. Now warm the paper in front of a fire, or over a gas flame, and observe that the writing begins to show, and appears blue. The pink salt has now lost water of crystallization, and turns blue. If the paper be left exposed to the moisture in the air, or more quickly if breathed upon, the salt will re-hydrate itself and turn pink again.

Some salts lose their water of crystallization on mere exposure to the air. Common washing soda is an example. If crystals of this are left on the table they soon lose their clear appearance and begin to fall to powder. This process is called *efflorescence.*

Other salts do just the opposite. They absorb moisture from the air; sometimes enough to cause the salt to liquify. Such salts are said to *deliquesce.* Substances which have this power are very useful to chemists for removing moisture from gases which are required to be dry.

HYDROGEN PEROXIDE.

Besides water, hydrogen and oxygen form another compound called hydrogen peroxide. It is obtained by acting on either sodium peroxide or barium peroxide with dilute sulphuric acid. It is a very unstable substance, being easily converted into oxygen and water. The readiness with which it parts with oxygen enables it to oxidize other substances. Thus, if it be added to lead sulphide (a black compound), it converts it into lead sulphate (a white substance). It is used on this account to clean old oil paintings which have become black by the white lead in the paint being changed to lead sulphide. Hydrogen peroxide also has bleaching properties, and is sometimes employed to bleach the hair.

EPITOME.

Water is a compound of oxygen and hydrogen. This is proved synthetically (1) by burning hydrogen in oxygen, and collecting

and testing the liquid product that is formed. (2) By exploding
oxygen and hydrogen in a closed vessel (Cavendish's method).
(3) By reducing certain metallic oxides, such as copper oxide, in
a stream of hydrogen (Duma's method). It may be proved
analytically (1) by decomposing water by certain metals, as sodium,
potassium, magnesium, or iron; whereby oxides of the metals
are formed and hydrogen liberated. (2) By heating steam with
chlorine, whereby hydrogen chloride (hydrochloric acid) is formed
and oxygen set free. (3) By the electrolytic decomposition of
water, whereby both oxygen and hydrogen are liberated.

Both by synthesis and analysis, we learn that water contains
its two constituents combined in the proportion of one part by
weight of hydrogen, to eight parts by weight of oxygen; and also
that the proportion by volume is two volumes of hydrogen to one
volume of oxygen.

When two volumes of hydrogen and one volume of oxygen
combine, they give two volumes of steam. When seen in mass,
water appears bluish-green. Its point of maximum density is 4° C.
When it freezes it expands $\frac{1}{10}$ of its volume.

CHAPTER IX.

As soon as chemists began to study chemical changes from a quantitative point of view; that is to say, as soon as they realized that certain relations existed between the *quantities* of the substances which take part in chemical changes, it then became important to be able to measure and to weigh with great accuracy; and as chemistry has become more and more an *exact science,* the various methods of measuring and weighing have developed in accuracy and delicacy.

The Metric System.—The measures and weights used in all scientific work are those of the French metric system. It is called the *metric* system because the standard unit of length is the *metre.* This corresponds roughly to the English yard, being 39·37 inches, and to this measure the standards of *capacity* and of *weight* are very simply related.

(1) *Measures of Length.*—The metre is divided into tenths, hundredths, and thousandths, which are called respectively, *decimetres, centimetres,* and *millimetres.* There are therefore 10 millimetres in 1 centimetre, and 10 centimetres in 1 decimetre.

The relation between these subdivisions and the English inch will be seen by Fig. 58, which shows side by side a decimetre scale, divided into ten centimetres, and each of these again into ten millimetres ; and a four-inch scale, divided into sixteenths. Roughly speaking, 2·5 cm., or 25 mm., are equal to 1 inch.

For the most part, the chemist uses only one measure of

length, viz. the millimetre; he records a length as 20 mm., or 760 mm., as the case may be, instead of 2 cm., or 7 dm. 6 cm.

1000 metres, or 1 kilometre, is a little over half a mile (0·62 miles), and is the unit employed on the continent of Europe for measuring distances. Instead of *milestones*, kilometre stones are employed upon the roads.

(2) *Measures of Capacity, or Volume.*—The space occupied by a cube whose sides measure one centimetre (*see* dotted cube, Fig. 58); that is to say, the volume of *one cubic centimetre* is taken as the unit. As there are 10 centimetres in a decimetre, there will be 10 × 10 × 10 = 1000 *cubic* centimetres in a *cubic* decimetre.

A cubic decimetre is called a *litre*; a litre, therefore, contains 1000 cubic centimetres. (The litre is equal to about 1¾ pints, and is the common continental unit of measure for liquids.)

(3) *Measures of Weight.*—The weight of one cubic centimetre (1 cc.) of pure water (measured at a temperature when water is most dense, see p. 79) is taken as the standard unit of weight, and is called a *gramme* (spelt in English, *gram*). Fig. 59 represents a gram weight (in brass), exact size; it is equal to about 15½ grains; and 31·1 grams are equal to 1 oz. Troy weight. The gram is subdivided into tenths, hundredths, and thousandths, called *decigrams*, *centigrams*, and *milligrams* respectively.

The kilogram (frequently called simply a "kilo") is 1000 grams. It is equal to about 2¼ lbs., and is the common continental unit.

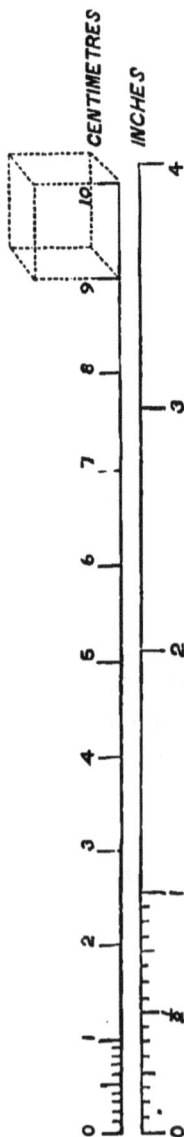

FIG. 58.

FIG. 59.

Since 1 cc. of water weighs 1 gram, it is obvious that we at once know the weight of any volume of water expressed in cubic centimetres; or, *vice versâ*, we know the volume of any weight of water given in grams. For instance, 1000 cc. of water—that is, 1 litre—weighs 1000 grams, or 1 kilo.

Similarly, if we know the specific gravity of a liquid (that is, how many times lighter or heavier than water it is) we can easily find out the weight in grams of any volume of it expressed in cubic centimetres, or *vice versâ*. For example, suppose we have 200 cc. of a liquid whose specific gravity is 0·5, and we wish to know how much it weighs :—

$$\text{Then, as } 1 : 0\text{·}5 :: 200 : x,$$

$$x = \frac{200 \times 0\text{·}5}{1} = 100 \text{ grams.}$$

Or, again, the specific gravity of mercury is 13·6, what will be the volume of 408 grams of this liquid?

$$\text{Then, as } 13\text{·}6 : 1 :: 408 : x,$$

$$x = \frac{408 \times 1}{13\text{·}6} = 30 \text{ cc.}$$

Instruments for Weighing and Measuring.—(1) **Thermometers.** These are instruments for measuring temperature. Their use depends upon the fact that liquids expand when warmed, and contract when cooled. The liquid most commonly employed is either mercury or alcohol. Water would not be suitable, because when moderately cooled it freezes.[1] The graduations, or *degrees*, upon the stem of a thermometer are purely arbitrary divisions. The instrument is first placed in a vessel filled with broken ice, and the position of the liquid in the stem is marked upon the glass. It is then placed in steam, from water which is kept briskly boiling. The liquid in the thermometer at once expands and rises in the stem, and the point to which it reaches is also marked upon the glass. These, then, are the two fixed points of the scale, namely, the temperature of melting ice and of steam from boiling water. The space upon the stem between these two fixed points is afterwards divided into a certain

[1] The student must refer to text-books on physics for details of the method of making and graduating thermometers.

number of equal parts, and the divisions continued both above and below. It is quite optional how many equal divisions the space between the two fixed points shall be divided into. For instance, we might decide to divide it into ten equal parts, calling the lower starting point o. Then, on our scale, the melting point of ice would be $0°$; the boiling point of water would be $10°$; blood heat would be $3\cdot7°$; and the ordinary temperature of a warm room would be $1\cdot5°$; but it will be obvious that it would be best to adopt one recognized scale. Unfortunately there are three scales in common use. In one, the space between the fixed points is divided into 80 equal parts, the lower starting point being the zero. In the second, it is divided into 100 equal parts, the zero being the same. In the third, it is divided into 108 equal parts, and the zero is placed 32 divisions *below* the lower fixed point.

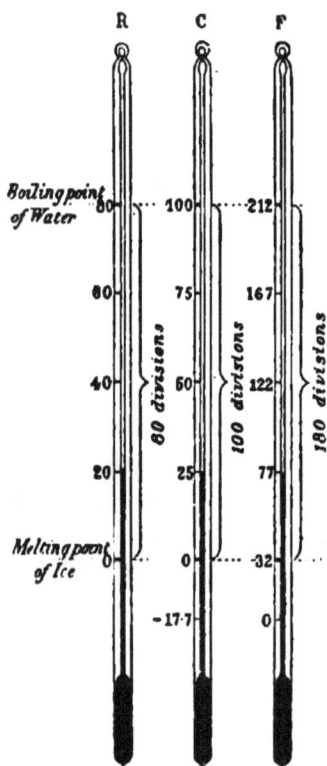

FIG. 60.

The first of these, called the *Réaumur* thermometer (R. Fig. 60), is the one in common use on the continent of Europe. The third is known as the *Fahrenheit* thermometer (F.), and is commonly used in England for ordinary purposes. As measured by this scale, the melting point of ice (or the freezing point of water) is $32°$, and the boiling point of water is $212°$. The other is the *Celsius*, or more commonly known as the *Centigrade* (C.) thermometer, and is always used for scientific purposes.

The relation in which these three scales stand to each other will be evident from the figure. It is obvious that the divisions on the C. scale are nearly twice as long as those of the F. scale, 100 C. divisions being equal to 180 F. degrees, or $1° C. = 1\cdot8° F.$

In order, therefore, to translate temperatures given upon the centigrade scale to degrees Fahrenheit, it is only necessary to multiply by 1·8, and add 32, because the zero of the latter scale is placed 32 divisions below the point at which the centigrade scale starts. And *vice versâ*, to convert degrees F. into degrees C., we must first subtract 32 and then divide by 1·8 :—

$$(n° \text{ C.} \times 1\text{·}8) + 32 = °\text{F.} \; ; \text{ and } \frac{n° \text{ F.} - 32}{1\text{·}8} = °\text{C.}$$

Example (1).[1]—The temperature of a bath is 95° C. What would this be on the F. scale?

$$95 \times 1\text{·}8 = 171\text{·}6 \; ; \qquad 171\text{·}6 + 32 = 203\text{·}6.$$
$$\therefore 95° \text{ C.} = 203\text{·}6° \text{ F.}$$

Example (2).—The temperature recorded on a hot summer day was 95° F. What is this on the C. scale?

$$95 - 32 = 63 \; ; \qquad \frac{63}{1\text{·}8} = 35.$$
$$\therefore 95° \text{ F.} = 35° \text{ C.}$$

(2) **The Barometer** is an instrument for measuring the pressure of the atmosphere. In its simplest form it consists of a long straight glass tube closed at one end, which has been filled completely with mercury, and inverted with its open end in a dish of mercury.

Experiment 85.—Take a stout glass tube about a metre long and seal up one end in the blow-pipe (see p. 37). Pour mercury into the tube until it is *nearly* full, firmly close the open end with the thumb, and tip the tube so as to allow a large air bubble to travel to the other end. This will sweep out all the small bubbles which were sticking to the glass. Now completely fill up the tube with mercury, once more close it with the thumb and invert it, and lower the end into a dish or trough containing mercury and carefully withdraw the thumb. Notice that the mercury falls in the tube to a certain point, and there remains stationary (*a*, Fig. 61).

---

[1] The student should set himself a few such examples to work out. For instance, he may verify the various temperatures which are shown to correspond on the scales in Fig. 60.

Now, why does the mercury fall in the tube at all? and why, if it falls at all, does it not drop down altogether? In order to answer these questions let us first inquire what there is in the clear space above the mercury. Can it be *air?* Unless a little air was accidentally allowed to get in when the tube was inverted in the mercury dish, it can hardly be air, because the tube was entirely filled with mercury, and the thumb was not removed until after the mouth of the tube was actually under the surface of the liquid. Try the following experiment :—

Experiment 86.—Slowly tilt the tube, and notice that the mercury goes up nearer and nearer to the top, until at last the tube is as completely full of mercury as it was before it was put into the trough. This proves that there is no air or any other gas in the tube. Raise it once more to the perpendicular position and the mercury again falls in the tube.

This experiment shows that there is *nothing* in the space above the mercury, but that it is a vacuum.[1] This being the case, it follows that if we were to connect one end of a long tube to the most perfect air-pump, and dip the lower end in mercury, it would be impossible to pump the mercury up the tube to a greater height than it stands in the tube in the barometer, because in that tube there is a perfect vacuum in the space above the mercury.

Experiment 87.—Measure the height of the column of mercury in the tube, measuring from the surface of the liquid in the dish to the top of the arched surface (called the *maniscus*), *a*, Fig. 61, of that in the tube. It will be found to be about 760 mm. [If no millimetre measure is at hand, use a foot rule, which will give nearly 30 inches as the height.]

Now stand the apparatus against a wall, and make a mark opposite the top of the mercury in the tube, and rule a horizontal line along the wall at that height. Then gradually tilt the tube into the positions *t, t'*, Fig. 61, and notice that all the time, the top of the mercury keeps level with the horizontal line, that is to say, it keeps at exactly the same actual vertical height.

[1] This experiment was first made by the Italian physicist *Torricelli*, and the space is called the *Torricellian Vacuum.*

The mercury is kept up in the tube, because the atmosphere pressing down upon the mercury in the dish, pushes it up until the weight of liquid in the tube exactly balances the pressure of the atmosphere. If from any causes the pressure

of the atmosphere be-comes greater, it will push the mercury higher up in the tube; and, in the same way, if the at-mospheric pressure be-comes less, it cannot sup-port so long a column of mercury, and therefore the liquid sinks a little in the tube. It is per-fectly easy to show by experiment, that it is the pressure of the air upon the liquid in the dish which keeps the mercury up in the tube. If, for instance, we surround

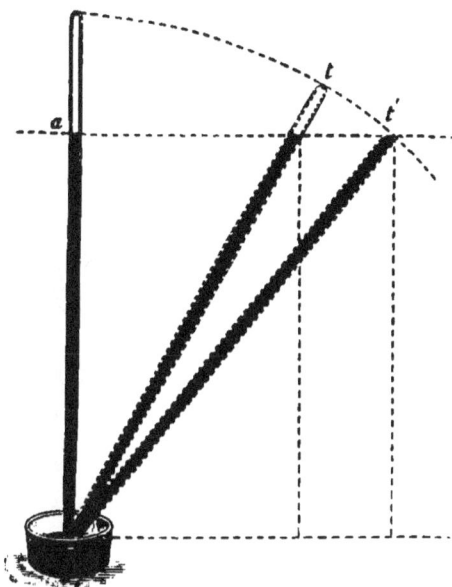

FIG. 61.

the apparatus with a tall glass shade, and with an air-pump gradually pump the air out of the shade, we should notice that the mercury would gradually sink lower and lower in the tube. By the time we had pumped out half the air (that is, reduced the pressure of the air on the mercury in the dish to one half), we should find that the mercury was now standing at only half its original height in the tube, namely, 380 mm. instead of 760.

Again, if we were to break open the top of the long tube, the mercury would of course instantly sink down to the same level as that in the dish, because the atmospheric pressure would then be the same on the liquid inside and outside the tube.

The pressure of the atmosphere is different at different parts of the earth's surface; and is also liable to vary in the same locality, from hour to hour. Consequently the height of the column of mercury which it is able to support also varies; therefore the barometer enables us to tell at any moment what is the actual atmospheric pressure at the time. What is called the *standard* or *normal* pressure is that which is able to support a column of mercury 760 mm. high.

(3) **The Balance.**—In order to weigh any substance accurately, or to detect minute differences in the weights of different things, it is necessary to employ a delicate balance and exact weights. For instance, a few grains of sand sprinkled upon the scale of an ordinary kitchen balance will

FIG. 62.

make absolutely no difference, whereas a single particle of sand placed upon one pan of a chemical balance would completely weigh it down. The chemical balance is usually enclosed in a glass case, partly to protect it from dust and dirt, and partly in order that it may not be exposed to the slightest draught when being used. Such a balance will readily weigh to a fraction of a milligram.

For the purposes of the elementary student, an instrument of the extremest delicacy is not necessary or desirable. Fig 62 shows a balance suitable for his requirements.

By turning the handle or lever H, the beam is liberated from its support, and is then free to swing. This balance will turn with 2 milligrams, and is able to carry as much as 100 grams. It must not be used for heavier weights than this.

The set of weights consists of the following, 50, 20, 10, 10, 5, 2, 1, 1, 1, grams, making up 100 grams; and 0·5, 0·2, 0·1, 0·1, 0·05, 0·02, 0·01, 0·01.

In order to make a weighing, we proceed as follows—

Experiment 88.—Place the object to be weighed, say a clean porcelain crucible with its lid, upon the left pan, and put on the other pan a weight which is roughly judged to be equal to it, say the 20-gram weight. Use the forceps (not the fingers) to lift the weights, and place them gently upon the scale pan. Now release the beam, by means of the lever, and observe which scale pan falls; we will suppose the 20-gram weight is too much. Raise the beam again, lift off the weight, *return it to its place in the box*, and put the 10-gram weight on the pan. Suppose this is too little, put on the 5-gram in addition. The weights must never be removed or added while the beam is free, but only when it is supported in the rest. If 15 grams is too much, remove the 5, return it to its place, and put on the 2-gram: if this is too little put on 1 gram, and if this is too much then the crucible weighs *more* than 12 but *less* than 13 grams. Now add 0·5 gram; suppose this to be too heavy, remove it and try 0·2. If this is too little add 0·1, if now too much remove the 0·1 and put on 0·05. Suppose this to be the exact weight, then, as the beam oscillates, the pointer will swing to the same distance upon the scale on both sides, and the weight of the crucible is 12·25 grams. [NOTE.—*When not in use, the balance and the weights should not be left exposed to the laboratory atmosphere, but should be either covered with a glass shade, or put away into a cupboard.*]

(4) **For measuring liquids,** graduated glass vessels are employed. For moderate volumes, a graduated 250 cc. cylinder may be used (Fig. 63). While for small quantities a *burette* is more useful (Fig. 64).

When small definite volumes are required, say exactly 10 cc. or 25 cc., a *pipette* is employed. This is dipped into the liquid, which is then sucked up the tube, nearly to the top, by the mouth, and the top quickly closed with the finger (Fig 65). By cautiously releasing the finger, the liquid is allowed to drop out until it sinks exactly to the mark on the narrow stem. When larger definite volumes are required, flasks holding ¼, ½, or 1 litre are used.

Experiment 89.—Select a flask with rather a narrow neck, and of such a size that when 500 cc. of water have been measured into it (by twice filling the 250 cc. cylinder) the water stands in the neck of the flask. Now gum a label round the neck in such a position that the top edge of the label is exactly level with the

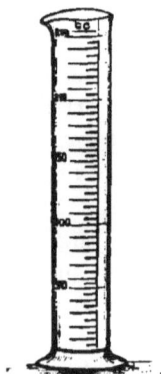

FIG. 63.　　　　FIG. 64.　　　　FIG. 65.

water. If now, by means of a file, a slight scratch be made upon the glass, just where the upper edge of the label is, a half litre measuring flask will have been made.

The volume which a given weight of a liquid occupies, however, depends upon the *temperature* of the liquid.

Experiment 90.—Fill the half litre flask exactly to the mark with cold water; place the flask upon a piece of wire gauze upon a tripod and gradually warm it with a gas flame. Notice that the water rises in the neck of the flask some distance above the mark.

The water has *expanded*, or increased in bulk. Now, by means of a pipette, withdraw some of the water, until exactly the half litre is left, and let the flask cool again. When it is cold, observe that it no longer contains as much as half a litre of water.

This shows that, if we measure a volume of liquid at a low temperature, we have a greater actual *weight* of it than if we measure it at a high temperature, and therefore we ought to measure always at the same temperature, or else make a proper allowance for the expansion or contraction. In ordinary experiments, when extreme accuracy is not necessary, and the quantities are only small, measurements made at the ordinary temperature of the room are regarded as being made at the same temperature, and no correction is made. The wholesale spirit dealer, on the other hand, takes careful note of the temperature at which his spirit is measured, else if he were to buy large volumes in hot weather, and sell it in cold weather, he would be a considerable loser by the transaction.

(5) **Measuring Gases.**—We may measure the volume of a gas obtained in a chemical process, by collecting it over water or mercury in a vessel graduated into cubic centimetres ; or we can collect it in an ungraduated vessel, mark the volume by gumming a label upon the glass, and afterwards seeing what volume of water has to be poured into the vessel from the 250 cc. measure, to fill it up to the mark. This is a comparatively rough method ; in more exact work it is usual to measure gases in glass tubes over mercury by means of .a millimetre scale. Vessels used for measuring gases are called *eudiometers*. Gases expand when warmed, and contract on being cooled just as liquids do, but to a much greater extent.

Experiment 91.—Fit a cork and delivery tube into a dry half-litre flask, and arrange it as shown in Fig. 66. Gently heat the empty flask with a lamp, and notice that the air within the flask quickly expands, and escaping through the delivery tube, is collected in the cylinder over water. Observe that the volume of the gas which is driven out of the flask is much greater than the volume of liquid which was removed from the flask of water in Exp. 90 by the pipette. Remove the flame, and notice that, as the air in the flask cools, and therefore contracts, water is drawn back from the trough to take its place.

When measuring the volume of gases, therefore, it is always

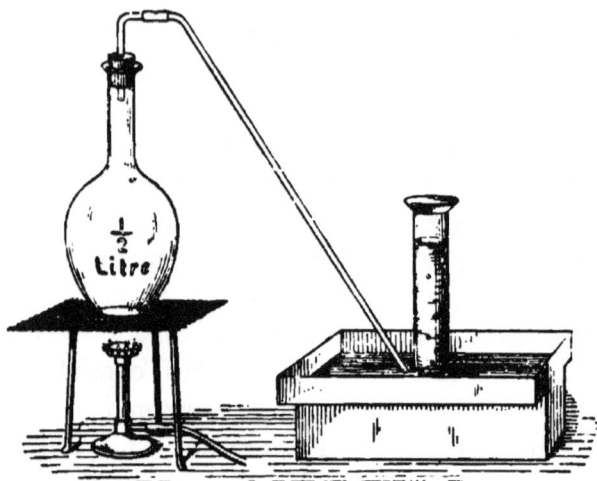

FIG. 66.

necessary to take into account the temperature of the gas at the time, and then to calculate what the observed volume would be if the temperature were to be lowered to 0° C., which is taken as the standard or normal temperature.

CHAPTER X.

SOME GENERAL PROPERTIES OF GASES.

Relation of the Volume of Gases to Heat.—Nearly a century ago it was discovered by Charles and Gay-Lussac that *all gases expand and contract to the same extent under the same changes of temperature, provided there is no alteration in the pressure.* This is known as the Law of Charles.

For every degree of the centigrade scale that a gas is heated, it expands $\frac{1}{273}$ part of the volume it occupies at $0°$ C.

Thus, 1 vol. at $0°$ C. becomes $1 + \frac{1}{273}$ at $1°$, and $1 + \frac{2}{273}$ at $2°$, and so on.

Or 273 vols. at 0 become $273 + 1$, or 274 at $1°$
,,   ,,   ,,   ,, $273 + 2$, ,, 275 ,, $2°$
,,   ,,   ,,   ,, $273 + 3$, ,, 276 ,, $3°$, etc.
,,   ,, $t°$, $273 + t$,   ,, $t°$

If $V°$ stands for the volume at $0°$ C., and V for the volume at t temperature, then we get the simple proportion—

$$273 : 273 + t :: V° : V,$$

or, $V = \dfrac{V° \times 273 + t}{273}$; and $\qquad V° = \dfrac{V \times 273}{273 + t}$

EXAMPLE 1.—Suppose we have a quantity of gas which measures 320 cc., while the temperature is $15°$ C. What volume will this occupy at $0°$?

$$\text{Then } V° = \frac{320 \times 273}{273 + 15} = 303\cdot3 \text{ cc.}$$

EXAMPLE 2.—250 cc. of gas measured at $-10°$ C. What will be the volume at the normal temperature?

$$\text{Then } V° = \frac{250 \times 273}{273 - 10} = 259\cdot5$$

H

EXAMPLE 3.—A quantity of gas occupies 100 cc. when measured at 0°. What will it measure when heated to 30°?

$$\text{Then } V = \frac{100 \times (273 + 30)}{273} = 110 \cdot 9 \text{ cc.}$$

EXAMPLE 4.—560 volumes of gas measured at 10°, are heated to 20°. How many volumes will they then occupy?

$$\text{Then, } 273 + 10 : 273 + 20 :: 560 : V$$

$$\text{or, } V = \frac{560 \times (273 + 20)}{273 + 10} = 576 \text{ volumes}$$

Relation of the Volume of Gases to Pressure.—If

we squeeze a solid or a liquid, we observe no change in their volumes; but if we put pressure upon a gas, it at once becomes greatly reduced in volume. Robert Boyle (1661) discovered that *the volume occupied by a given weight of any gas is inversely as the pressure.* This is known as Boyle's Law. It means that if we double the pressure on a gas we reduce the volume to one-half; while if we diminish the pressure to one-half, we increase the volume to double. It is easy to put this law to the test. Suppose a quantity of air is enclosed in the tube T (Fig. 67), which is connected by means of a thick india-rubber tube to a small reservoir of mercury, capable of being hoisted up and down. Let us first place the reservoir in such a position that the mercury in it is exactly level with that in the tube. Under these circumstances the enclosed air is under the ordinary atmospheric

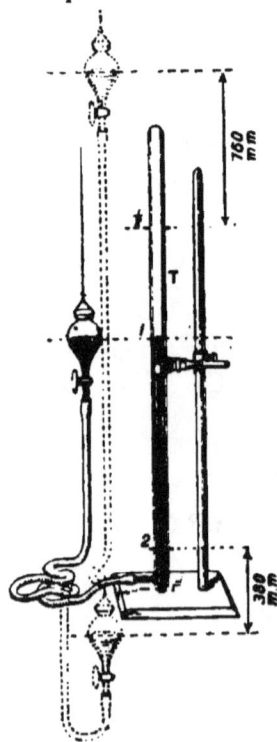

FIG. 67.

pressure. Let us call it 1 vol., and mark it upon the tube. Next we will place a mark at an equal distance below, and another at one-half the distance above it.

Now suppose we raise the mercury reservoir, we shall see the volume of the gas diminishing until presently the mercury stands at the $\frac{1}{2}$-mark. When it reaches this point we shall find, on measuring, that the height of the mercury in the reservoir above that in the tube is about 760 mm. We have already learnt that 760 m.n. of mercury represents a pressure equal to that of the atmosphere, hence we have subjected the gas within the tube to a pressure of an additional atmosphere; that is, there is now twice the pressure upon it that there was at first, and its volume is reduced to one-half.

In the same way, if the reservoir be lowered down until the gas in the tube has expanded to twice its original vol., that is, down to the 2 mark, we should find, on measuring, that the mercury in the reservoir was standing 380 mm. *below* that in the tube. Instead of being under the ordinary atmospheric pressure, the gas is now under *reduced* pressure. It is subjected to a pressure of 760 mm. − 380 mm. = 380 mm. That is, to a pressure of only *half* the ordinary atmospheric pressure, and we see, as demanded by Boyle's law, that its volume is doubled.

Experiment 92.—Take a glass tube about half a metre long, and closed at one end, and completely fill it with mercury. Now pour out this mercury into a cc. measure, and in this way ascertain the exact capacity of the tube. Then pour back into the tube one-half the mercury, and mark upon the tube where it comes to. This mark will therefore indicate exactly half the capacity of the tube. Now close the tube with the thumb and invert it in a dish of mercury, and stand it in a vertical position. Notice that the mercury does not stand at the mark, but at a point some distance below it. Why is this, for the tube when inverted was half filled with air and half with mercury? The reason is because the gas is under *reduced pressure*, and has therefore expanded. It is under a pressure equal to that of the atmosphere *minus* that of the column of mercury in the tube. We must, therefore, first ascertain the actual height of the barometer at the time, and then carefully measure the height of the column of mercury, from the surface of that in the dish to the top of that in the tube. Suppose the barometer happens to be low, say 740 mm., and the length of the column in the tube to be 200 mm., then 740 − 200 = 540 mm. is the pressure to which the gas is exposed.

Experiment 93.—Take the tube used above, fill it with *water* up to the mark and invert it in a dish of water. Notice where the

water stands ; it only falls a very little way below the mark.   Why,
is this?   It is because, water being so much lighter than mercury
(13½ times lighter), the short column of it in the tube scarcely
reduces the pressure on the gas at all.

Seeing that the volume of a gas is so closely dependent
upon the pressure to which the gas is exposed, it will be
evident that if we wish to compare volumes of gases, we must
know what they measure under the same conditions of pressure.   Now it is not often possible, and is seldom convenient,
to actually put gases under one regular pressure before measuring their volumes, so the plan always adopted is to measure
the volume of the gas at the particular pressure it happens to
be under, and then to calculate what volume it would occupy
at a pressure of 760 mm., this being the standard or normal
pressure.   This is called *correcting for pressure*, and the calculation is a simple proportion, based on Boyle's law.

For example, 100 cc. of a gas measured at 380 mm., what
volume would it occupy at 760 mm. ?

$$\text{Then, as } 760 : 380 :: 100 : x$$
$$x = \frac{100 \times 380}{760} = 50 \text{ cc.}$$

Again, a quantity of gas measured 500 volumes at 780 mm.
What will it occupy at the standard pressure ?

$$\text{As } 760 : 780 :: 500 : x$$
$$x = \frac{500 \times 780}{760} = 513 \text{ volumes.}$$

In practice, the corrections for both temperature (p. 97)
and pressure are usually made together.   For example,[1] a
sample of gas measured 360 cc, its temperature was 15° C.,
and it was at atmospheric pressure ; but the barometer at the
time was standing at 750 mm.

$$\text{Then } \frac{360 \times 273}{273 + 15} = \text{the temperature correction alone,}$$
$$\text{and } \frac{360 \times 750}{760} = \text{the pressure correction alone.}$$

[1] The student should make himself perfectly familiar with this method
of reducing volumes of gas to the normal temperature and pressure, by
working out a number of examples.

Putting the two together, we get—

$$\frac{360 \times 273 \times 750}{(273 + 15) \times 760} = \frac{73710000}{218880} = 336\cdot7 \text{ cc.}$$

Therefore, 360 cc. of gas at 15° C. and 750 mm. $=$ 336·7 cc. at the normal temperature and pressure.

The Crith.—One litre of hydrogen gas, measured at 0° C. and 760 mm., weighs 0·0896 grams. This is an extremely important figure to remember, for by means of it we can calculate the weight of any volume of any gas, so long as we know the *density* of that gas; that is, how many times heavier than hydrogen it is. For instance, if we learnt that oxygen was sixteen times as heavy a gas as hydrogen, bulk for bulk, then 1 litre of oxygen will, of course, weigh sixteen times as much as 1 litre of hydrogen; that is, 0·0896 × 16, or 1·42 grams, at normal temperature and pressure (N.T.P.). This number, 0·0896, is so important that a name has been given to it (just as in mathematics the name of the Greek letter π is used to denote the number 3·1416, the ratio between the diameter and the circumference of a circle). The name adopted for the weight of a litre of hydrogen measured at N.T.P. is the Greek word signifying a barleycorn, *crith*. It is used symbolically for a *little weight*, and does not mean that a litre of hydrogen weighs as much as a barleycorn. We say that a litre of oxygen weighs 16 criths ; that is, simply sixteen times 0·0896. Another important number to be remembered in this connection is the volume of 1 gram of hydrogen, measured at N.T.P. 1 gram of hydrogen measures 11·165 litres at N.T.P. From this number, just as from the *crith*, we are able to calculate the weight of any volume of any gas whose density is known. Thus, if oxygen is sixteen times as heavy as hydrogen, then, since 1 gram of hydrogen measures 11·165 litres, obviously, 16 grams of oxygen will measure 11·165 litres. In other words, 11·165 litres is the volume which will be occupied by the same number of grams of a gas, as expresses the density of that gas. Suppose we have three gases whose densities are respectively 14, 18½, and 22 (that is, one is 14 times, one 18½ times, and the other 22 times as heavy as hydrogen), then 14 grams of one, 18¼ grams of the second, and 22 grams of the third will measure 11·165 litres.

The liquefaction of gases.—We have learnt (p. 97) that when a gas is heated from 0° to 1° it expands by $\frac{1}{273}$ of its bulk ; if it be cooled from 0° to −1° it *contracts* $\frac{1}{273}$ of its volume; from 0° to −2, −3, −4, etc., it contracts $\frac{2}{273}$, $\frac{3}{273}$, and $\frac{4}{273}$ respectively. Now, if this law holds good, however much we cool a gas, it would follow that if a quantity of gas be cooled from 0° to −273°, it would contract $\frac{273}{273}$ of its volume, that is, it would occupy *no* volume at all.

Now, a temperature as low as −273° has never yet been reached (this temperature is sometimes called the *absolute zero*). The lowest degree of cold which has been obtained as yet, is about −220°, but before this point is reached all the known gases except one, and that one is hydrogen, pass into the liquid condition. Just as steam, when cooled, changes from the gaseous state to the liquid, so all other gases, when sufficiently strongly cooled, change from the gaseous to the liquid condition. Some gases require very little cooling to make them do this, while others require to be exposed to the lowest possible temperature in order to make them change their state. Among the latter class are oxygen and nitrogen, and it is because it is only recently that chemists have been able to obtain the necessary degree of cold, that these gases have only of late years been obtained in the liquid state. For example, oxygen requires to be cooled down to the extremely low temperature of −181° to cause it to pass from gas to liquid.

There is little doubt but that before long a sufficiently low temperature will have been reached in order to produce liquid hydrogen.

As a gas gets near to the temperature at which it turns into a liquid, it begins to depart from the law of Charles and Gay-Lussac, and, of course, when it liquefies, and therefore ceases to be a gas, it is no longer subject to the laws which govern gases.

Again, it is found that when gases are subjected to pressure, they sooner or later begin to depart from Boyle's law, and finally to change their state from gases to liquids. A most remarkable point, however, about the effect of pressure in causing the liquefaction of a gas, is, that the gas *must* be below a certain temperature. If it be above this temperature, no amount of pressure will squeeze it into the liquid state. This particular temperature is different for each gas, and is called the *critical temperature* of the gas. For example, the critical temperature of chlorine is 141°. This means that if chlorine is heated above this point, no amount of pressure will make it pass into the liquid state ; but at all ordinary temperatures of the air, chlorine is *far* below its critical temperature, therefore it can be easily condensed to the liquid state by mere pressure.

Again, the critical temperature of ethylene (see p. 184) is about 10°, that is, just a trifle below the ordinary temperature of a room ; hence, in order to compress this gas into the liquid state, it must be slightly cooled, so as to bring its temperature below its critical point.

In the case of some gases the critical temperature is extremely low, thus in the case of oxygen it is −118·8°. Therefore, in order to compress oxygen into a liquid, it is absolutely necessary to cool it down to this intense degree of cold before liquefaction will take place. At −118·8° oxygen requires a pressure of fifty atmospheres to liquefy it. The more it is cooled below this, the less pressure is needed to liquefy it, until at −181° it passes into the liquid condition at the ordinary atmospheric pressure.

The critical temperature of hydrogen is lower still, and up to the present no artificial cold has been obtained low enough to cool hydrogen in quantity down to this point ; hence hydrogen has not yet been obtained as a coherent liquid, although by special devices momentary indications of liquidity have been observed, when the hydrogen appeared as a froth or spray.

CHAPTER XI.

SIMPLE QUANTITATIVE MANIPULATIONS.[1]

EVERY one knows that if we mix two substances together, say salt and sugar, we can have the two ingredients present in any proportion we please. We could, for instance, make a mixture containing so large a proportion of salt, that the sugar present could not be tasted; or the sugar might so preponderate that we could not taste the salt in the mixture. Now a most important question arises, viz. Can substances enter into chemical combination in this manner? That is to say, when two elements unite, can they do so in any proportion? does the composition of the compound they produce vary? If two substances can combine together in any proportion, just as they can be mixed together, then obviously the composition of the compound produced will depend upon the quantities of the ingredients that were employed to produce it. If, on the other hand, it should be that there is some fixed proportion in which elements combine, that if they unite it must be in some particular proportions, or not at all, then it would follow that any given compound would always have absolutely the same composition. This is a point which it is of the very greatest importance to settle, and in order to do so we must make experiments upon weighed or measured quantities of substances, and carefully weigh or measure the resulting products.

---

[1] The quantitative experiments have been arranged as far as possible so as to make use only of such knowledge of chemical facts as the student has already gained from the earlier chapters. The teacher will do well to substitute or add others from time to time. It is well that the student should check his own results by doing duplicate experiments.

In other words, we must make *quantitative* experiments, as distinguished from *qualitative.*

We can make our quantitative investigations in two ways: either synthetically or analytically. We can cause weighed quantities of materials to enter into chemical union, and weigh the products; or we can decompose known quantities of certain compounds, and weigh or measure the resulting substances.

Experiment 94.—The combination of magnesium with oxygen. We have learnt that when substances burn in the air, they are combining with oxygen to produce oxides. Let us now burn a weighed quantity of magnesium and weigh the resulting oxide. Take a porcelain crucible with its cover, and counterpoise it on the balance (p. 92). Instead of using the weights, place a pill-box in the opposite pan, and put into it some fine shot, until the crucible is exactly balanced. This little box of shot will now be the "tare" for the empty crucible.

Next weigh out into the crucible from half to three-quarters of a gram of magnesium ribbon, which has been previously scraped or rubbed down with sandpaper to remove any oxide from it, and cut into short pieces. Weigh it carefully, and note the weight.[1]

Now place the crucible upon a pipe-clay triangle supported on a tripod or the ring of a retort stand, and heat it (Fig. 68); applying the heat first gently and then more strongly. From time to time slightly raise the cover by means of tongs, to observe the

FIG. 68.

progress of the combustion, but not so as to allow any of the oxide to be carried away.

When the whole of the magnesium has burnt, allow the crucible to cool. Then place it on the balance, put the "tare" on the opposite scale, and weigh the contents of the crucible. The first

---

[1] Instead of using a tare, the crucible may be weighed, first empty, and again with the metal in it; then, by deducting the first from the second weight, the weight of magnesium is found. The student will find it quicker, and will be less likely to introduce errors, if he uses tares as described.

thing observed will be that it has gained in weight. This we have learned to expect from Exps. 72 and 73. Carefully note the weight.

The difference between the weight of the magnesium employed, and that of the oxide obtained, is the weight of oxygen from the atmosphere which has combined with the magnesium.

Calculate the proportions in the following way :—

Suppose the weight of magnesium used was 0·53 gram, and that of the magnesium oxide produced to have been 0·88 gram ; then, subtracting the former from the latter, we get—

$$0·88$$
$$0·53$$
$$\overline{}$$
$$0·35 = \text{weight of oxygen}$$

which has combined with 0·53 grains of magnesium.

Then, as 0·35 : 0·53 :: 1 : x

$$x = \frac{0·53 \times 1}{0·35} = 1·5$$

therefore the proportion in which oxygen and magnesium have combined together in this experiment is 1 part by weight of oxygen with 1·5 parts of magnesium.

In this experiment the magnesium, being heated in the air, is surrounded with an unlimited supply of oxygen; therefore, if we were to repeat the process a number of times, each time employing a different weight of the metal, and were to find that in every case the magnesium and oxygen combined in exactly the same proportions, we should have good ground for supposing that these two elements could only unite in some definite proportion.

It will at this point be interesting to find out whether the same result would follow if the magnesium were to be burnt in pure oxygen, instead of in the air. For this purpose the next experiment should be made.

Experiment 95.—Take a piece of combustion tube (p. 32) with moderately thin walls, about 18 cm. (7 inches) long, and slightly border the ends (see p. 36). Into one end fit a cork with a short straight glass tube, and insert into the other a short plug of cotton wool, as shown in Figure 69. Now counterpoise the whole as in

Exp. 94, with a pill-box of fine shot. Then remove the cork and place in the tube some cleaned fragments of magnesium ribbon (rather less than half a gram). Replace the cork and carefully weigh the magnesium by adding weights to the "tare."

Support the tube on a stand as in the figure, and pass a slow stream of oxygen through it ; the gas being previously collected in a gas holder, or being used from a cylinder of compressed oxygen. In order to see the rate at which the gas is passing, it should be

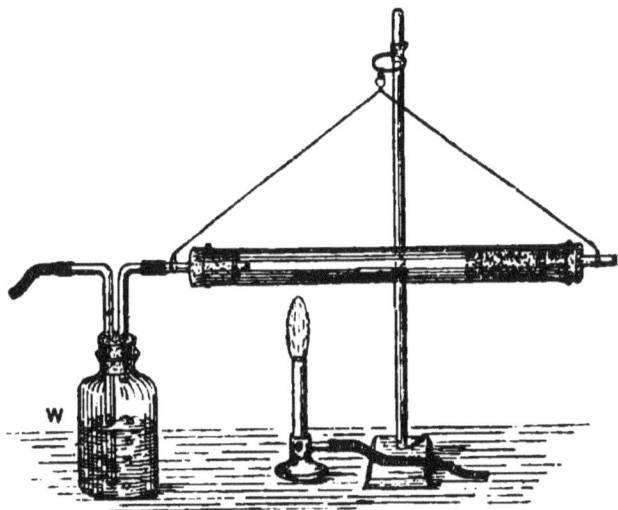

FIG. 69.

made to bubble through water in the manner shown at W in the figure. Now heat the magnesium, applying the heat gradually at first, and then holding the flame so as to ignite the metal nearest to the cork. The metal burns with great brightness, and if the gas is not passing in too fast, the plug of wool will prevent any of the white oxide from escaping. When all the metal is burnt, and the tube has cooled, again weigh it, and calculate the proportions in which the two elements have combined, as in the former experiment. If the result is the same, then we have not only proved that combustion in air is the same process as combustion in pure oxygen, but we have strengthened our suspicion that chemical combination takes place only between fixed proportions of the elements.

Instead of repeating these experiments a great many times, in order to prove beyond all doubt that the elements magnesium and oxygen always combine in the same fixed

proportions, it will be better to experiment with other substances, and see if they also show indications of a similar
behaviour.

Experiment 96.—**The combination of sulphur with oxygen.**
Take a piece of combustion tube as in Exp. 95, and fit a cork and
tube into each end.   Fit up a U tube with corks and bent tubes as
shown in Fig. 70, attaching a thin wire to it by which it can be
suspended.   Place in this tube a small quantity of a solution of
sodium hydroxide, so as just to fill the bend.

Now counterpoise the combustion tube with shot, then place a
small fragment of sulphur (from $\frac{1}{2}$ to $\frac{3}{4}$ gram) in the tube and
carefully weigh it.   Note the weight, and remove the "tare" from

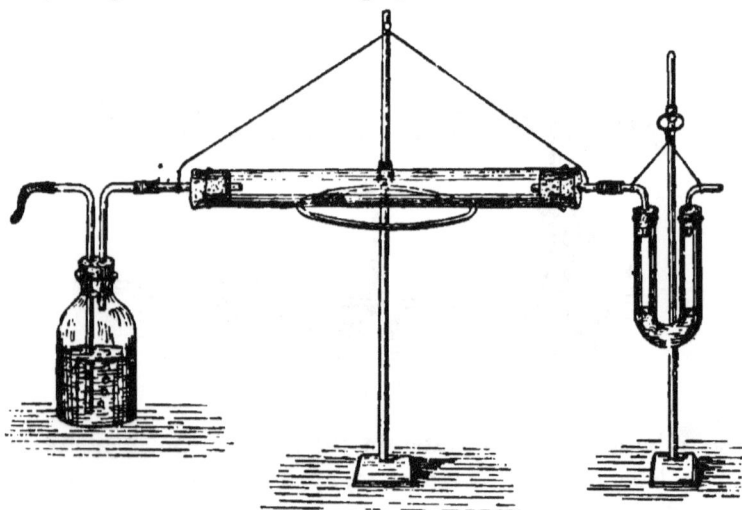

FIG. 70.

the balance, placing it upon a clean sheet of paper upon which can
be notified what the tare belongs to.

Next suspend the U tube from the hook at the end of the arm
of the balance, and counterpoise this with another tare, carefully
preserving the latter along with the former one.

The U tube is now to be attached to the combustion tube, as
shown in the figure, and a *slow* stream of oxygen passed through
the apparatus using the bottle W, as in Exp. 95.   By comparing
the rate at which the gas bubbles through the wash-bottle and the
U tube, it will be easy to see whether the connections are all tight.
Now gently heat the sulphur.   It, of course, first melts, and hence
it is necessary that the tube shall be quite horizontal.   Presently
it takes fire and burns, producing *sulphur dioxide*, the same

compound as was formed in Exp. 64. This oxide, as we then saw, is a gas, and in order to catch it, we are now making it bubble through the solution of sodium hydroxide in the U tube. This readily *absorbs* sulphur dioxide. Notice that as soon as the sulphur begins to burn, the bubbling in the U tube gets slower and (provided the stream of oxygen is not too fast) soon stops altogether, although it continues bubbling through the bottle at the same rate as before. By the time the sulphur has burnt away, however, it will be seen that gas once more passes out through the U tube. Continue the stream of oxygen for a minute or two after the sulphur is extinguished, in order to be sure that all the gaseous oxide has been driven into the U tube, and then stop the experiment.

Now disconnect the U tube, hang it on the balance and place its tare in the other scale. Observe that the apparatus has gained weight : find how much by adding weights. Note the weight. Next place the combustion tube on the balance, with its own tare, and see if it is still counterpoised. If it is, then the whole of the sulphur has been burnt ; but if not, add weights to find the exact weight of what is left in the tube, and note this weight. We have now got the weight of sulphur that has been burnt, and the weight of the oxide of sulphur that has been produced.

For example—

Suppose the weight of sulphur put into tube = 0·62 grams
And weight of residual sulphur left in tube = 0·01   „

Then the actual weight of sulphur burnt = 0·61 grams

And suppose the increase of weight of the U tube = 1·22 grams ; then, subtracting the weight of sulphur burnt from the weight of sulphur dioxide obtained, we get 1·22
0·61

0·61 grams = weight of oxygen

which has combined with 0·61 grams of sulphur. Therefore the two elements oxygen and sulphur have combined together in equal proportions, 1 part of oxygen to 1 part of sulphur.

Just as in the case of magnesium and oxygen, if we were to repeat this experiment a number of times, and each time get the same result, we should be forced to the conclusion that sulphur and oxygen do not unite together in *any* proportions, but that there is some fixed relation between the weights of each which enter into combination.

If we repeat this experiment, using air (contained in a gas-holder) instead of oxygen, we shall be able to confirm the former conclusion we have come to (p. 107), viz. that the same products of combustion are formed whether air or pure oxygen be employed, and we shall, at the same time, get a second result to serve as a check upon the first.

In carrying out this experiment, it will be instructive to notice that, during the burning of the sulphur, there is scarcely any perceptible change in the rate at which bubbles pass through the liquid in the U tube. This is because in the air the oxygen is mixed with a large excess (four times as much) of nitrogen, which takes no part in the experiment, but simply passes through the apparatus unaffected. Both in the case of magnesium and oxygen, and sulphur and oxygen, we have been dealing with instances of the simple union of two elements; let us take one more illustration of the synthetical formation of a compound, but this time produced by *indirect* processes.

We have learnt from Exp. 80 (p. 74) that hydrogen and oxygen can be made to combine indirectly by first uniting the oxygen to copper (forming copper oxide), and then reducing this compound by means of hydrogen, which deprives the copper oxide of oxygen, leaving metallic copper. We shall now repeat this experiment quantitatively.

Experiment 97.—**The combination of hydrogen with oxygen.** Prepare a piece of combustion tube, as in Exp. 95, with a cork and straight tube fitted into one end, and place in the tube a roll of clean copper gauze (one of the bright rolls after Exp. 80 will answer best). Heat the tube containing the copper, with a Bunsen flame, and, while it is hot, pass a stream of oxygen through it, so as to convert a portion of the metal into its oxide. Then allow the tube to cool, place it with its contents upon the balance, and counterpoise it with shot. Carefully preserve this tare. Now attach to the other end of the tube the little collecting apparatus, A, Fig. 71. This consists of a test-tube fitted with a cork and two bent tubes. The longer one is drawn out to a thin taper and reaches nearly to the bottom of the tube, while its other end passes through a cork which will properly fit the end of the combustion tube. Strong sulphuric acid is poured into the test-tube, to the depth of about one third, care being taken that the acid does not wet the neck of the tube where the cork fits, and the apparatus is then suspended by a wire upon the balance, and counterpoised with shot ; the tare being carefully preserved. As soon as it has been counterpoised, a little cap should be slipped upon the end E of the short tube, in order to prevent moisture from the air getting in when it is not being used. (The cap is made by means of a short piece of indiarubber tube, with a piece of glass rod pushed into one end, as is shown at C, Fig. 71.) When this is attached to the combustion tube, the little cap is removed, and a slow stream of hydrogen is passed through

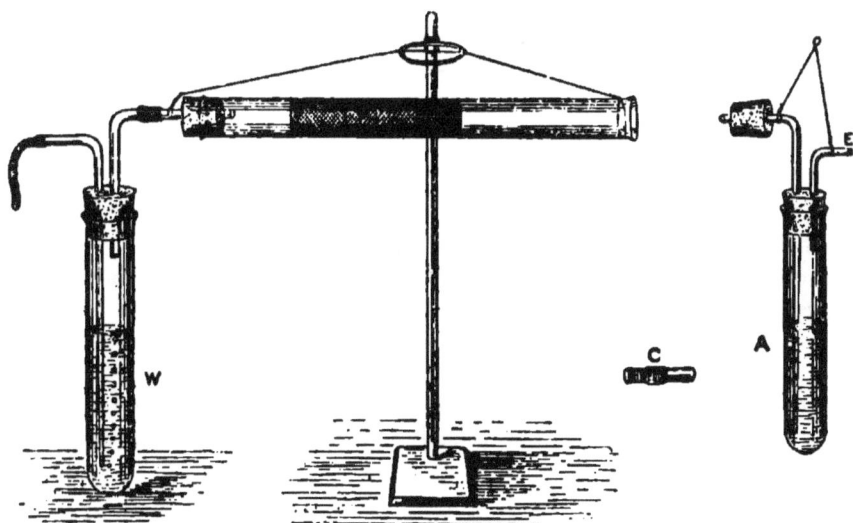

FIG. 71.

the apparatus ; the tube at the same time being strongly heated. As what we are ultimately to collect and weigh is *water*, it is very necessary that the hydrogen we use should not bring any moisture into the apparatus, or this would spoil the result ; the hydrogen, therefore, *before* it enters the tube containing the copper oxide must be made to pass first through an arrangement (*w*, Fig. 71) similar to the one we are using to absorb the water formed in the experiment, where it bubbles through strong sulphuric acid, from a fine drawn out tube. When the copper in the tube appears bright again, remove the lamp, and allow the apparatus to cool, while the hydrogen still slowly passes through. Then stop the gas, disconnect the collecting apparatus, and place it on the balance with its own tare. It is of course heavier than before, on account of the water it has absorbed. Carefully weigh the increase, and note down the weight. Now place the combustion tube on the balance, with its tare; notice that it has *lost* in weight. Find how much loss by putting weights in the scale containing the tube. Note this weight. The *gain* in weight of the collecting tube, is the weight of water formed in the experiment. The *loss* of weight suffered by the combustion tube, is the oxygen which has been taken away from the copper oxide.

Suppose the weight of water formed was 1·25 grams, and the weight of oxygen from the copper oxide 1·11 grams; then, subtracting the weight of oxygen from the weight of water, we get

1·25
1·11
———

0·14 grams = weight of hydrogen which has combined with 1·11 grams of oxygen.

$$\text{Then, as } 0\text{·}14 : 1\text{·}11 :: 1 : x$$
$$x = \frac{1\text{·}11 \times 1}{0\text{·}14} = 7\text{·}9$$

Therefore hydrogen and oxygen have combined together in the proportion of 1 part of hydrogen to 7·9 (or nearly 8) parts of oxygen.

This experiment has been made a number of times by different chemists, with much more care and a great many more precautions against possible errors. Fig. 72 shows the

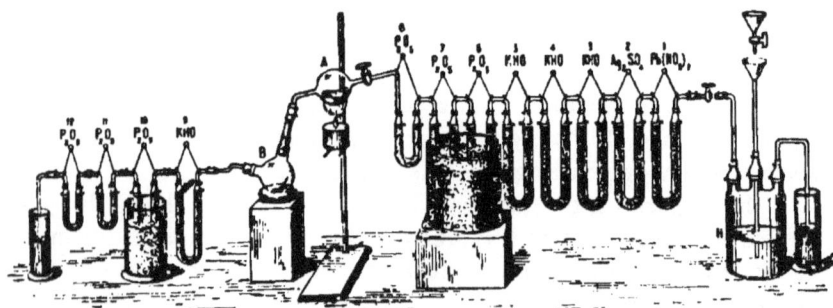

FIG. 72.

famous apparatus employed by Dumas (1843), who made many very exact determinations of the proportion by weight in which oxygen and hydrogen combine. In essence it is the same as that which has been used in the last experiment. The weighed copper oxide was heated in the bulb A. Hydrogen was generated in the bottle H, and the water formed was collected in the bulb B. The U tubes 1 to 8 were to remove all impurities from the hydrogen, and render it absolutely dry; tubes 9, 10, 11, were to absorb any water vapour which did not condense in the bulb B, and were therefore weighed before and after the experiment, and any increase in weight was added to the weight of the water in B. Tube 12 was merely a guard tube, to prevent any moisture from the air getting into tube 11.

If it is true that elements can combine together only in some fixed proportions, it must follow that any particular compound will always have the same composition. Let us now test the matter from this point, and ascertain whether the same compound always does contain its constituent elements in the same proportions.

Experiment 98.—The decomposition of potassium chlorate. Exp. 59 has taught us that this compound contains oxygen as one of its constituent elements ; and that when it is heated it gives up this oxygen, and that a residue of potassium chloride is left.

Counterpoise a porcelain crucible with its lid, and weigh into it a small quantity (say 2 or 3 grams) of potassium chlorate. Heat the crucible, as in Exp. 92, keeping the lid on, because the salt decrepitates (that is, " crackles ") when first heated, and some might be lost if the lid were off. When all effervescence is at an end, the crucible is allowed to cool, and is then weighed. The loss of weight is the oxygen which has been given up.

Suppose the weight of potassium chlorate used to have been 2·34 grams, and the weight of the residue after heating to have been 1·42 grams, then the difference, namely, 0·92 grams = weight of oxygen evolved, therefore 2·34 grams of potassium chlorate were made up of—

$$\begin{aligned} \text{Potassium chloride} &= 1\cdot42 \text{ grams} \\ \text{oxygen} &= 0\cdot92 \quad ,, \\ \hline &\ 2\cdot34 \end{aligned}$$

or, calculating as in the former cases, we get the proportion—

oxygen 1 part to potassium chloride 1·54 parts.

If on repeating this experiment, using various samples of

I

potassium chlorate, the same proportion is always found to exist, we should infer that this compound always has the same composition.

How to calculate the Percentage Composition.—Sometimes chemists prefer to express the results of such experiments as the foregoing, in parts per hundred; that is to say, to state how much of the different ingredients there are in 100 parts of the compound. This *percentage composition* of a compound is calculated from the experimental numbers in the following way, taking the last experiment as an example.

If 2·34 grams of potassium chlorate contain 1·42 grams of potassium chloride, what will 100 parts contain?

$$\text{As } 2\cdot34 : 100 :: 1\cdot42 : x$$
$$x = \frac{1\cdot42 \times 100}{2\cdot34} = 60\cdot7$$

Again, if 2·34 grams of potassium chlorate contain 0·92 grams of oxygen, what will 100 parts contain?

$$\text{As } 2\cdot34 : 100 :: 0\cdot92 : x$$
$$x = \frac{0\cdot92 \times 100}{2\cdot34} = 39\cdot3$$

Therefore the percentage composition of potassium chlorate, as calculated from the results obtained in Exp. 98,[1] is—

$$\text{Potassium chloride} = 60\cdot7$$
$$\text{oxygen} = 39\cdot3$$
$$\overline{100\cdot0}$$

Experiment 99.—**The decomposition of mercuric oxide.** Counterpoise a hard glass test-tube, and weigh into it a small quantity of mercuric oxide. Note the weight. Now heat the red oxide in the bottom of the tube, until it has entirely decomposed into oxygen (which escapes), and mercury which sublimes upon the upper part of the tube; that is, until nothing is left in the

[1] The student should calculate in a similar manner, from his results obtained in the preceding experiments, the percentage composition of magnesium oxide, of sulphur dioxide, and of water.

bottom of the tube. When cold, weigh again; this gives the weight of the mercury, and the difference between this and the former weight is the oxygen which has been expelled. For example—

Suppose the weight of mercuric oxide to have been 5·40 grams, and the weight of mercury left in the tube 5·00 grams, then 5·40

$$\begin{array}{r} 5\cdot00 \\ \hline 0\cdot40 \end{array}$$

0·40 grams = weight of oxygen which was expelled from 5·40 grams of the oxide.

Calculated as in the former cases, we find that the proportion of oxygen to mercury is as 1 to 12·5.

To express the percentage composition of the compound, we have—

$$5\cdot40 : 100 :: 5\cdot00 : x = 92\cdot59 \text{ mercury}$$
$$\text{and } 5\cdot40 : 100 :: 0\cdot40 : x = 7\cdot41 \text{ oxygen}$$
$$\overline{100\cdot00}$$

In both of the last experiments we have ascertained the amount of oxygen by *difference,* assuming the loss suffered on heating the compounds to be entirely oxygen. It will be obvious, however, that we must be quite sure that nothing else but oxygen is expelled by heating. Suppose, for instance, that either the potassium chlorate or the mercuric oxide contained some moisture; this would be driven off along with the oxygen, and the loss of weight due to this would be counted as oxygen. We can check the former result by modifying the experiment so as to collect and measure the oxygen, and then by the methods of calculation explained on page 101, we can find from the volume of gas collected what its weight is, without actually weighing it.

Experiment 100.—Fit a short, narrow, hard glass test-tube (or piece of combustion tube closed at one end) with a cork and narrow delivery tube. Counterpoise the empty tube, and weigh into it a small quantity of mercuric oxide. Attach the delivery tube, and proceed to heat the compound, collecting the oxygen over water in the usual way in a cylinder, capable of holding about a litre. As soon as the whole of the oxide is decomposed, disconnect the delivery tube, and, when cold, again weigh the tube, to find the weight of the mercury. Now mark the volume of gas in the cylinder [1] by sticking a strip of a gummed label upon the glass. Take the temperature of the water in the trough by dipping a thermometer into it; this may be taken as the temperature of the gas. To find the exact pressure under which the gas is, we must measure the height of the water in the cylinder above the level of the liquid in the trough, and divide this by 13·6 (because mercury is 13·6 times as heavy as water); and then subtract the result from the height of the barometer at the time. (If, however, the cylinder used is not a long one, and if the atmospheric pressure is not very far from the normal, viz. 760 mm., the final result will not be much influenced by this correction, and it can be omitted.)

Now remove the cylinder from the trough, and empty it. Then pour water into it from a graduated cc. measure,[1] until the liquid reaches to the edge of the paper mark. The number of cc.s required to do this will give the volume of oxygen, measured at the observed temperature and pressure.

In the following example, correction has been made for pressure as well as temperature.

Weight of mercuric oxide taken = 8·90 grams
„   „   mercury remaining = 8·24   „
Number of cc.s of water required to fill the cylinder
up to the label . . . . } = 485 cc.
That is, the volume of oxygen as collected . .
Temperature = 16° C.
Pressure—Height of water in cylinder, above level in trough
= 90 millimetres
Corresponds to $\frac{90}{13·6}$ = 6·6 millimetres of mercury
Barometer reading = 772·6
∴ Actual pressure = 772·6 − 6·6 = 766 mm.

(1) 485 cc. of oxygen, at 16° and 766 mm., what volume at N.T.P.?

$$\frac{485 \times 273 \times 766}{(273 + 16) \times 760} = 461·7 \text{ cc.}$$

(2) What is the weight of 461·7 cc. of oxygen?

1000 cc. (1 litre) of oxygen weighs 0·0896 × 16 = 1·4336 grams (see p. 101).

∴ 1000 : 461·7 :: 1·4336 : x = 0·66 grams.

[1] If a *graduated* cc. cylinder is used in which to collect the gas, the volume can be read off at once, and the operation of measuring the capacity of the cylinder will of course be unnecessary.

Therefore ·66 grams of oxygen were obtained from 8·90 grams of mercuric oxide. '

Or oxygen and mercury are present in the proportion of 1 : 12·5.

First Law of Chemical Combination.—By making a large number of experiments like these, we are driven to the general conclusion that elements do not combine in *any* proportion, but that they always unite in some definite and fixed quantities. We have in this way discovered a general truth, or a Law, as it is termed—the *Law of Constant Proportion*, which may be stated in the following words, *The same compound always contains the same elements, combined together in the same proportion by weight.*

Hydrogen and oxygen, we have seen, unite in the proportion by weight of 1 part of hydrogen to 8 parts of oxygen. If, therefore, these two gases be mixed together in any other proportion before combination, there will remain over after their union the excess of either the one or the other as the case may be. Thus, if equal weights were present in the mixture before chemical union, when combination took place the oxygen would still only combine with $\frac{1}{8}$th of its weight of the hydrogen, leaving the remaining $\frac{7}{8}$ths of the hydrogen unaffected. Similarly, if any excess of oxygen be present beyond the proportion of 1 part of hydrogen to 8 parts of oxygen, this excess remains over unaltered after combination of the two elements.

Oxygen and magnesium again combine in the proportion of 1 to 1·5. Any excess of either element beyond these proportions that may be present before combination, is left behind unaltered when the elements unite.

Second Law of Chemical Combination.—In the course of experimenting with a view to establish the law of constant proportion, chemists meet with a number of cases which at first appear like exceptions, for it is found that very often two elements do combine together in more than one proportion. Whenever this happens, however, the compounds that are produced are totally different from each other; and moreover the proportion in which the elements are present in each, is found to be always the same for the same compound;

that is to say, each compound has its own constant composition. For instance, we have found that when sulphur burns in air or in oxygen, an oxide of sulphur is formed which contains the two elements in the proportion S : O = 1 : 1, and we have learnt that the composition of this compound is always the same. But under certain conditions, sulphur and oxygen can be made to unite in a different proportion, namely, S : O = 1 : 1·5. The compound that is obtained under these circumstances, however, is a totally different substance from the other; its composition is always absolutely constant, never departing from the proportion of 1 part sulphur to $1\frac{1}{2}$ parts of oxygen. Similarly the element carbon combines with oxygen in two proportions, giving rise to two distinct compounds, one of them contains the elements in the proportion of carbon 1 part, and oxygen 1·33 parts, while the other is composed of carbon 1 part, and oxygen 2·66 parts; that is, it contains twice as large a proportion of oxygen. The composition of each compound never varies, and therefore they are not exceptions to the law of constant proportions.

Another striking example of the same elements combining in more than one proportion is the case of oxygen and nitrogen. These two elements unite in no less than *five* different proportions, giving five different compounds, each one quite different from the rest, and each having a perfectly constant composition. Their names and composition are as follows :—

Nitrous oxide	1 part of nitrogen to 0·57 parts of oxygen.			
Nitric oxide	1 ,,	,,	1·14 ,,	,,
Nitrogen trioxide ...	1 ,,	,,	1·71 ,,	,,
Nitrogen peroxide ...	1 ,,	,,	2·28 ,,	,,
Nitrogen pentoxide	1 ,,	,,	2·85 ,,	,,

That is, the second, third, fourth, and fifth of the compounds contain respectively twice, three times, four times, and five times as much oxygen in their composition as the first ; or the relative proportions of oxygen which are combined with a fixed quantity of nitrogen are as 1 : 2 : 3 : 4 : 5

In all cases like these, where the same elements combine in more than one proportion, forming more than one compound, we find the same simple ratio existing between the different weights of the element which can combine with a constant weight of the other. This is another general truth, or law, known as the *Law of Multiple Proportions*, which may be thus stated—

When the same two elements unite to form more than one compound, the different weights of the one, that combine with a constant weight of the other, bear a simple ratio to one another.

CHAPTER XII.

HYDROCHLORIC ACID.

In beginning the study of this substance, the first point to settle is whether it is an element or a compound, and if the latter, what elements it is composed of. We have already learnt (p. 47), that when this substance is acted upon by zinc, hydrogen is evolved. It must, therefore, contain hydrogen, and therefore it is a compound; but what is the hydrogen combined with? What else does the compound contain besides hydrogen?

The only other *compound* we have examined so far, is water; let us, therefore, try to decompose hydrochloric acid as water was decomposed, by an electric current.

Experiment 101.—Place some hydrochloric acid in the apparatus (Fig. 56, p. 76) and connect the wires to a battery. Collect all the gas in a short stout tube, as in Exp. 81. When the tube is full, apply a light to it. Notice that it explodes very much as the gas obtained from water did. Can it be oxygen and hydrogen?

Experiment 102.—Collect the gas from each wire in separate tubes, as in Exp. 82. Notice that the two gases are not evolved in the proportions they were when water was decomposed. In that case there was twice as much hydrogen as oxygen, here the two gases collect in nearly equal volumes.[1] Test the gas which collects from the wire which is connected with the zinc (or negative) plate of the battery, by removing the test-tube and applying a light to the gas. Note that it burns in the familiar manner of hydrogen. Notice that the gas in the other tube is not quite colourless, but is faintly yellow. Cautiously bring a lighted taper into the tube and

[1] If the current is passed through the acid for some little time before the gases are collected, they will collect in *equal* volumes.

observe the behaviour of the flame. It does not burn brilliantly, therefore the gas is certainly not oxygen. Neither is the flame altogether extinguished, but it burns with a strangely lurid and smoky manner.

Once more connect the battery for a few moments, and gently sniff the gas which is evolved, observe that it has a most irritating, cough-producing effect upon the air passages of the nose and upon the throat; quite different from the smell of the hydrochloric acid itself. This is a new gas, for neither oxygen nor hydrogen have any smell, or colour, or behave towards a taper-flame as this gas does. This yellowish, choking gas is called *chlorine;* the name being from the Greek word for yellowish green.

In the case of water, we found, when the two gases obtained by electrolysis were mixed together in a strong glass vessel and exploded, that a small quantity of liquid water was formed and that there was no gas left in the vessel (Cavendish's experiment). Let us repeat this experiment with the two gases given off when hydrochloric acid is submitted to electrolysis.

Experiment 103.—Fit a small bottle with a cork carrying a glass tube bent at a right angle. Push through the cork two brass (or, better, platinum) wires, the ends of which are twisted round two rods of carbon (as in Fig. 73) about the thickness of a slate pencil (such carbon rods as are used for electric lamps). Fill the bottle with hydrochloric acid nearly up to the wires, and connect the wires to a battery. Allow the current to pass for about ten minutes. Then connect to the exit-tube a long stout glass tube (Fig. 74), having a stop-cock at each end, and provided with two platinum wires sealed into the glass. Allow the gases to stream through this tube for ten to fifteen minutes (depending on the rate at which the hydrochloric

FIG. 73.   FIG. 74.

acid is being decomposed; that is, on the strength of battery being used), then close both cocks and stop the current. Now, by means of a Ruhmkorf coil, send an electric spark into the gases in the tube by the wires. The gases instantly explode, but notice that there is no noise, only a slight click accompanied by a flash of light in the tube. Carefully examine the tube to see if there is any indication of a liquid moistening the inside. There is none. If the hydrochloric acid which has been formed by the re-union of the gases has not condensed to a liquid, where is it? Dip one end of the tube beneath some mercury in a dish, and open the stop-cock at that end. Notice that the mercury is not sucked in, and also that no gas bubbles out. This shows that the tube is exactly full of gas, just as it was at first; therefore, if hydrochloric acid has been produced it must be a *gas* and not a liquid, and also it must occupy *exactly as much space or volume as the mixture of gases that were put into the tube.* Now open one end of the tube beneath water. Notice that the water is sucked up into the tube exactly as though the tube were vacuous!—but it was *not* vacuous, because, had it been so, the mercury would have entered before. Test a little of the water which has been drawn up into the tube, with a piece of litmus paper; notice that the water is *acid.* The hydrochloric acid in the gaseous state which was formed by the union of the gases in the tube has *dissolved in the water.*

We see from this, that hydrochloric acid is a *gas* at ordinary temperatures. What, then, is the liquid we have used, and which is generally called *hydrochloric acid.* This liquid is simply *a strong solution of the gas in water.* In Exp. 103 we have obtained a small quantity of the gas by direct synthesis. Let us now prepare a larger quantity of it by an indirect method, and study its properties. To produce it, we shall use sulphuric acid and sodium chloride. We know that the first of these contains hydrogen; the sodium chloride (common salt) contains chlorine combined with sodium. When these two compounds are brought together a chemical change takes place. The sodium of the sodium chloride displaces the hydrogen of the sulphuric acid; and this hydrogen then unites with the chlorine which the sodium has left.

Experiment 104.—Measure 55 cc. of strong sulphuric acid into a flask, and slowly pour upon it 40 cc. of water, constantly shaking the mixture. The liquids become extremely hot, and must be allowed to cool. Fit the flask with a cork carrying a straight tube which just dips into the liquid, and a short bent exit tube. Put a handful of common salt into the cold acid, attach a delivery tube and prepare to collect the gas in the usual way. Little action takes place in the flask in the cold, but on gently warming it a plentiful evolution of gas takes place. At first the air within the apparatus is expelled and bubbles through the water in the trough ; but notice that presently the bubbles get smaller and smaller, until scarcely a trace of gas escapes through the water. Why is this? Evidently gaseous hydrochloric acid cannot be collected over water. Now disconnect the delivery tube, and collect a cylinder of the gas by downward displacement (p. 27). Notice that the gas is quite colourless, but that when it escapes into the air it fumes strongly. Gently fan a whiff of the gas towards the face, so as to smell it ; and note that, although very pungent, it is quite different from chlorine. When the cylinder is full, cover it with a stout glass plate, and quickly invert it mouth downward in a trough of water. Observe that the water quickly rushes up into the cylinder, as though the latter were vacuous, showing how *extremely soluble* this gas is, and how utterly impossible it would be to collect it over water. Test the water in the cylinder with litmus, see that it is acid. Although this gas cannot be collected over water, it can be collected over *mercury*, in a mercury pneumatic trough, for it has no action on this metal.

Experiment 105.—Collect a second cylinder of gas. Dip a moist litmus paper into it ; notice that the gas is powerfully acid. Test the behaviour of the gas towards a flame, by bringing a lighted taper to the mouth of the jar. Observe that the gas does not behave either like hydrogen or like chlorine ; for it does not burn, and does not allow the taper to continue burning even with a smoky flame.

Let us now try by chemical processes to expel first the hydrogen, and then the chlorine, from this gaseous compound of these two elements. To displace the hydrogen we will employ the metal sodium.

Experiment 106.—Place a fragment of sodium in a hard glass bulb-tube (as in Fig. 38, p. 44), and pass a brisk stream of gaseous hydrochloric acid over it. Attach a delivery tube to the exit, and collect any gas over water. When the air has been swept out of the bulb tube, proceed to heat the sodium, which quickly melts and presently takes fire. The moment it begins to burn, gas at once collects in the vessel in the pneumatic trough. Test this gas with a lighted taper, and recognize hydrogen by the familiar flame. What has become of the other constituent, namely, the chlorine ? Break open the bulb, and with the little finger bring a small quantity of the white residue on to the tongue. By the familiar taste recognize *common salt*, that is, sodium chloride ; therefore the chlorine has combined with the sodium. The combustion of the metal was in reality its rapid combination with chlorine.

In order to liberate the chlorine from hydrochloric acid, we may employ certain compounds rich in *oxygen*, so that the hydrogen in the acid may unite with this to form water, and let the chlorine go free. Substances belonging to a class called *peroxides* will do this. We will use manganese peroxide (also called manganese dioxide).

Experiment 107.—Place a quantity of manganese dioxide in a moderately wide glass tube, supported in a horizontal position and fitted at each end with a cork and glass tube. To one end connect a delivery tube, leading to the pneumatic trough, and attach the other end to the apparatus generating the gaseous hydrochloric acid. As the gas passes through the tube, the manganese dioxide may be slightly warmed by just brushing a flame along the tube. Notice that a gas is again collected through the water in the trough. Observe also that moisture collects upon the inside of the horizontal tube. This is the product of the union of the hydrogen present in the hydrochloric acid with oxygen out of the manganese dioxide. Test the gas with a lighted taper and observe the same smoky flame as in Exp. 102. Cautiously whiff the gas towards the face and recognize the same suffocating smell.

By Exps. 102 and 103 we have learnt that hydrochloric acid contains half its own volume of each of its constituent elements. So far as the volume of the hydrogen is concerned (but not that of the chlorine) this may be confirmed in the following way. The U-shaped apparatus (Fig. 75) is filled with gaseous hydrochloric acid down to the second division, the mercury being level in the two limbs. We then fill up the open limb with *sodium amalgam* (this is a solution of sodium in mercury, made by pressing pieces of sodium beneath the surface of mercury in a mortar with the pestle. The sodium dissolves in the mercury with great energy, sometimes even taking fire). The tube is then closed with the thumb, and tipped up so as to make the gas pass round the bend and bubble up through the amalgam. The gas is passed through the amalgam once or twice, and then returned to the closed limb. While in contact with the sodium amalgam, the gas is decomposed. The chlorine unites with the sodium, forming sodium chloride, which appears as a white crust on the mercury and the sides of the tube, while the hydrogen is set free, just as in Exp. 106.

When the mercury in both limbs is once more levelled, by running away the excess by means of the side tap, it is seen that the residual gas occupies exactly *one half* the original volume. We can prove that this gas is hydrogen by again filling the open limb with mercury so as to drive the gas out at the stopcock at the top, and on applying a light the issuing gas will burn with the familiar hydrogen flame.

Chlorides. — Compounds of elements with chlorine are called chlorides. Metallic chlorides, for example, are the compounds of metals with chlorine. Many of them are produced by the direct union of the metals with chlorine. We may, however, also define chlorides as compounds in which the hydrogen of hydrochloric acid is replaced by a metal. Thus, when zinc is acted on by hydrochloric acid, the hydrogen of the acid is displaced and liberated as gas, while zinc chloride

FIG. 75.

is at the same time produced. Again, if certain oxides of metals are acted on by hydrochloric acid, the hydrogen is again displaced and a chloride of the metal produced; but in this case the hydrogen is *not liberated as gas*, but goes to the oxygen of the oxide to produce water. The same result is obtained when a metallic hydroxide is brought into contact with hydrochloric acid; a chloride of the metal and water are formed. Since hydrochloric acid is a compound of hydrogen with chlorine, it is often called *hydrogen chloride.*

Tests for Chlorides.—Hydrogen chloride and all other

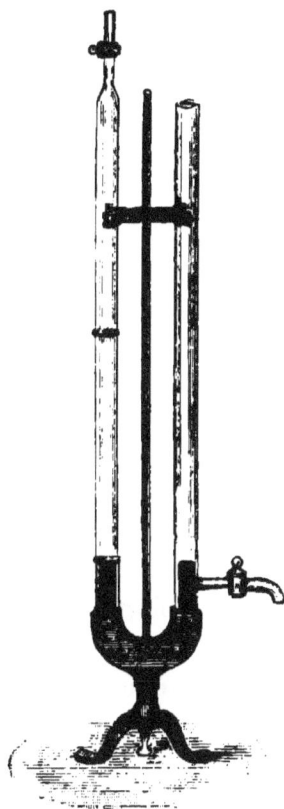

chlorides that are soluble in water, are detected by means of a solution of silver nitrate.

Experiment 108.—Dissolve a pinch of sodium chloride (common salt) in water, and add a few drops of silver nitrate solution. A white curdy precipitate, consisting of silver chloride, is at once produced. Divide the liquid containing the precipitate into two portions, and add some ammonia to one and nitric acid to the other. Note that the precipitate dissolves in ammonia, and does not dissolve in nitric acid. This distinguishes the precipitate from any other which would be produced by silver nitrate.

Lead nitrate also gives a similar looking precipitate when added to a solution containing a chloride, and this precipitate can be distinguished from others in the following way.

Experiment 109.—Add a few drops of a solution of lead nitrate to a solution of sodium chloride. Compare the precipitate of *lead chloride* with that obtained in the last experiment. Now boil the mixture, and notice that the precipitate disappears. It dissolves in hot water. Cool the test-tube, and observe white, silky-looking crystals begin to deposit. This is the lead chloride being deposited again in the form of crystals.

All chlorides, whether soluble in water or not, give off chlorine when warmed with a mixture of manganese dioxide and sulphuric acid, and can, therefore, be identified by this test, for the smell and colour of chlorine are unmistakable (see Chlorine, next chapter).

EPITOME.

Hydrochloric acid is a compound of hydrogen and chlorine. It is sometimes called hydrogen chloride. It can be produced synthetically by the direct union of the two constituent elements.

Hydrochloric acid is prepared by acting on common salt with sulphuric acid. Under ordinary circumstances it is a colourless gas, having a pungent acid smell. When it escapes into the air it fumes strongly. It is *extremely* soluble in water. At the ordinary temperature 1 litre of water will dissolve 450 litres of the gas. This solution of the gas in water is the common article of commerce known as hydrochloric acid (also sometimes called *muriatic acid*, and *spirit of salt*).

Gaseous hydrochloric acid is $1\frac{1}{4}$ times as heavy as air, and can therefore be collected by downward displacement. The gas does not burn, neither will it support the combustion of ordinary flames. It is strongly acid, but has no bleaching power.

The solution of the gas in water is decomposed by a current of electricity into equal volumes of hydrogen and chlorine ; and when these two gases in equal volumes unite, the gaseous hydrochloric acid which results occupies the same volume as the mixed gases did before their union.

Hydrochloric acid is manufactured on an enormous scale as an article of commerce. It is obtained by the same method as in the laboratory, namely, from salt and sulphuric acid, only the operation is conducted in enormous iron vessels, and the gas is made to pass at once into water, to obtain a solution of it.

Hydrochloric acid attacks many metals ; some, such as zinc and iron, very readily ; others, as copper and tin, less quickly. In all cases, hydrogen is given off, and a chloride of the metal is left. Hydrochloric acid has no action upon mercury, or upon the so-called " noble " metals, gold and platinum.

Gaseous hydrochloric acid is distinguished from all other gases by its extreme solubility in water, and by the formation of a white curdy precipitate of silver chloride, when a solution of silver nitrate is added to the solution of the gas ; the precipitate being insoluble in nitric acid, but easily soluble in ammonia. All soluble chlorides are tested for with silver nitrate in the same way.

Reactions for hydrochloric acid—

(1) Preparation from sodium chloride and sulphuric acid (labo· ratory process),

$$NaCl + H_2SO_4 = NaHSO_4 + HCl.$$

(2) Manufacturing process—when the mixture is more strongly heated, and *all* the hydrogen from the sulphuric acid is displaced by sodium—

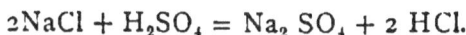

$$2NaCl + H_2SO_4 = Na_2 SO_4 + 2 HCl.$$

(3) Test for hydrochloric acid, with silver nitrate,

$$HCl + AgNO_3 = AgCl + HNO_3.$$

CHAPTER XIII

CHLORINE.[1]

From the last chapter we learn that chlorine is one of the constituent elements present in hydrochloric acid; and we saw in Exp. 107 how it may be obtained from this compound. When we require any quantity of this element, it is more convenient to start with a *solution* of hydrochloric acid instead of the gas.

Experiment 110.—Place a quantity of strong hydrochloric acid in a flask fitted with a cork and delivery tube, and add a handful of powdered manganese dioxide. Gently warm the flask, and collect the gas at the pneumatic trough, using strong brine instead of ordinary water. Collect several cylinders full. Notice the colour of the gas, which when seen in quantity, appears a greenish-yellow.

Instead of employing ready-made hydrochloric acid as the source of chlorine, we can use the materials from which hydrochloric acid is obtained, namely, common salt and sulphuric acid.

Experiment 111.—Place a quantity of sulphuric acid (of the same strength as that used in Exp. 104) in a flask, and add common salt and manganese dioxide in about equal parts. Apply a gentle heat, and collect the gas over strong salt-water, as above.

---

[1] In experimenting with chlorine, the student must be careful not to inhale the gas. Even when mixed with a large volume of air, it causes a most unpleasant irritation to the throat and lungs; and if a strong whiff of the gas is inhaled it becomes a dangerous poison. These experiments, therefore, are best performed in a good draught cupboard.

Experiment 112.—**Properties of Chlorine.** Take one of the jars of gas and place it mouth downward in ordinary water. Notice that the water slowly rises in the jar, and if allowed to stand some time will gradually creep up to the top. This shows that the gas is moderately soluble in water (contrast this with the case of hydrochloric acid). It is because chlorine is thus dissolved by water, that in Exp. 99 the volumes of the two gases were not quite equal. There was rather less chlorine collected than hydrogen. If when the water has risen a little, the jar be closed with a cork, and briskly shaken, the water will dissolve the gas more quickly, and it will be seen that the liquid has much the same yellowish colour as the gas. This solution is called *chlorine water*, and smells strongly of chlorine. Pour it into a stoppered bottle, and stand it in the window. If the sun happens to be shining on the window, notice that the yellow colour rapidly disappears, and at the same time the smell of the chlorine has gone. The same result follows more slowly in dull daylight. The chlorine, under the influence of light, decomposes the water in which it is dissolved, combining with the hydrogen to form hydrochloric acid, while oxygen from the water is set free.

We have seen in Exp. 103 that chlorine and hydrogen combine with explosion when a spark is passed into a mixture of these gases. The union of these elements also takes place if the mixture is exposed to daylight. It takes place slowly and noiselessly in dull daylight, but instantly with explosion in direct sunlight or in the bright light of burning magnesium.

Experiment 113.—Dip a moistened blue litmus paper into a jar of chlorine. Notice that instead of being reddened, as with hydrochloric acid, the paper is *bleached.* Do the same with some other colours, such as carmine, or aniline blue or violet, by staining pieces of paper with the dyes, and dipping them into the gas.

This bleaching power of chlorine is one of its most characteristic and important properties.

Experiment 114.—Collect a stoppered jar of chlorine by downward displacement, having first placed a small layer of strong sulphuric acid in the jar. Close the jar with a *cork*, and spread the acid over the interior as much as possible by tipping the vessel about. The strong acid has the power of removing any vapour of water which is present in the gas. Now dry a strip of paper which has been stained with carmine or an aniline colour, by holding it for a few moments in front of a fire, or over a gas flame. Then, with a touch of wax, attach one end of the paper to the stopper of the jar ; grease the stopper with vaseline, and as quickly as possible remove the cork and replace it by the stopper with the paper suspended from it. Leave it for some time, and notice that the colour is not bleached. Afterwards remove the stopper, allow a drop or two of water to touch the paper, and replace it in the jar. Note that the colour is instantly discharged from the moistened parts of the paper.

This shows that *dry* chlorine is incapable of bleaching : that the presence of water is necessary to the action. This is because the bleaching action of chlorine is really a process of *oxidation*. The chlorine decomposes the water (just as in Exp. 108), liberating oxygen, and uniting with the hydrogen to form hydrochloric acid (which we have learnt, from Exp. 103, has no bleaching power). Although we know that oxygen under ordinary circumstances is not able to bleach colours, at the moment it is liberated from combination it is endowed with a chemical activity which it has not got at other times. Elements in this condition are said to be in the *nascent state* (that is, the newly born state). Chemists frequently take advantage of the extra active nature of elements at the moment of their liberation, to bring about chemical changes which they are incapable of under ordinary conditions. In the case before us, the nascent oxygen oxidizes the colouring matter, and converts it into colourless compounds.

By former experiments we have found that chlorine does not burn, and that a taper burns in the gas with a very smoky flame. The wax of the taper is a compound of hydrogen and carbon, and from the manner in which it burns in chlorine, we might suppose that the combustion was due to the combination of the hydrogen with chlorine ; and as the carbon was set free in the form of soot, that this element does not readily unite with chlorine. Let us test this idea by a few experiments.

Experiment 115.—Lower a burning jet of coal-gas into chlorine. Notice that the flame continues burning, but throws off a quantity of carbon, as smoke, and also that fumes are produced similar to those which we see when hydrochloric acid escapes into the air.

Experiment 116.—Boil a little turpentine in a test-tube, and pour some of it on a piece of blotting-paper, and at once drop it into a jar of chlorine. The turpentine instantly bursts into flame, again throwing out a dense black cloud of carbon, and forming fumes of hydrochloric acid. Both coal-gas and turpentine are compounds of hydrogen and carbon.

Experiment 117.—Lower a burning jet of hydrogen into chlorine. Note that the flame produces no black smoke, because no carbon is present, but gives the same white fume of hydrochloric acid.

Experiment 118.—Unscrew the cup from a deflagrating spoon, and fasten a piece of charcoal to the rod with a piece of wire. Make the charcoal red-hot in a flame and quickly plunge it into chlorine. Notice that the charcoal does not burn, but is instantly extinguished. Chlorine will not combine directly with carbon, therefore it is, that in the foregoing experiments this element is rejected by the chlorine, and thrown out of combination.

From these experiments we see that, just as the bleaching action of chlorine is the outcome of its affinity for hydrogen, so its peculiar behaviour towards ordinary combustibles is due to the same cause. Indeed, these two elements are so eager to combine, that it is only necessary to expose a mixture of them to the influence of daylight to bring about their union. In dull daylight the combination takes place slowly, but if the mixture be exposed for a single moment to strong sunshine the two gases combine with an explosion. In whatever way the two elements combine, hydrochloric acid is in all cases the product of their union. Besides its powerful affinity for hydrogen, chlorine combines energetically with most metals. In many cases, if the metal is in the form of powder, or of thin leaf, the combination is so rapid that the metal actually takes fire.

Experiment 119.—Throw into a jar of chlorine a small quantity of finely divided iron (obtained by reducing the oxide in a stream of hydrogen), or a little finely powdered antimony. Or thrust into the jar a bundle of leaves of *Dutch metal* (that is, brass) tied to the end of a stout wire. In each case the metal takes fire in the chlorine, forming a chloride of the particular metal.

On account of the action of chlorine on metals, this gas cannot be collected over mercury, for it instantly attacks this metal.

EPITOME.

Chlorine is prepared by acting on hydrochloric acid with manganese dioxide, or by acting on a mixture of sodium chloride and sulphuric acid with manganese dioxide.

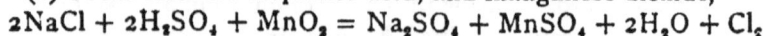

Chlorine is a greenish-yellow, suffocating, poisonous gas; two and a half times as heavy as air. It is moderately soluble in water (1 litre of water dissolves about 3 litres of chlorine at the ordinary temperature), the solution having the colour and smell of the gas.

Chlorine does not burn, but ordinary combustibles will burn in the gas with a smoky flame.

It combines readily with hydrogen, producing hydrochloric acid, and with metals, forming chlorides; in many cases the metal takes fire in the gas.

Chlorine is a powerful bleaching agent. This is its most important property, and enormous quantities of chlorine are manufactured by the action of manganese dioxide upon hydrochloric acid for this purpose. As chlorine in the free state is not a convenient article of commerce, the gas as it is produced is made to combine with slaked lime, which gives a compound known as *bleaching powder*. This substance readily gives up its chlorine when acted on by acids; therefore, if a piece of coloured material be dipped in a solution of bleaching powder, and then into a bath of dilute acid, chlorine is liberated in the pores of the fibre, and this quickly bleaches the colour.

Chlorine may be distinguished from all other gases by its colour and its bleaching property.

Reactions for chlorine—

(1) From hydrochloric acid and manganese dioxide,

$$4HCl + MnO_2 = MnCl_2 + 2H_2O + Cl_2$$

(2) From salt and sulphuric acid, and manganese dioxide,
$$2NaCl + 2H_2SO_4 + MnO_2 = Na_2SO_4 + MnSO_4 + 2H_2O + Cl_2$$

(3) Combination with hydrogen, $Cl_2 + H_2 = 2HCl$

(4) Combination with metals, (*a*) Sodium, $Na + Cl = NaCl$

(*b*) Copper, $Cu + Cl_2 = CuCl_2$

(*c*) Antimony, $Sb + 3Cl = SbCl_3$

Formation of bleaching powder,

$$CaH_2O_2 + Cl_2 = CaOCl_2 + H_2O$$

Action of acids on bleaching powder,

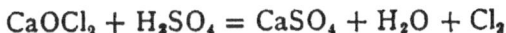

$$CaOCl_2 + H_2SO_4 = CaSO_4 + H_2O + Cl_2$$
$$CaOCl_2 + 2HCl = CaCl_2 + H_2O + Cl_2$$

The Halogens.—Chlorine is one of a family of four elements, which bear a close relation to each other in their chemical behaviour, although they are not much alike to look at.

These four elements are *fluorine* (a gas), *chlorine* (a gas), *bromine* (a liquid), and *iodine* (a solid). They are called the *halogens*.

Fluorine.—This element is extremely difficult to obtain, because it is so intensely active, that the moment it is set free from one combination it enters into another, even combining with the materials of almost any vessel in which the experiment is made.

The commonest natural compound of fluorine is *fluor spar*, that is, calcium fluoride.

The element itself is prepared by passing an electric current through a solution of potassium fluoride in perfectly pure hydro-fluoric acid. The experiment is conducted in apparatus constructed of platinum, which resists the action of fluorine better than anything else. Glass cannot be used at all, as even hydrofluoric acid acts readily on this substance.

Bromine is a heavy deep-red liquid, which readily passes off into vapour of the same colour. It has a strong suffocating smell (the name bromine signifies a *stench*), and attacks the nose and throat, as chlorine does. It also irritates the eyes.

In nature it occurs combined with sodium and with magnesium in the salt obtained from sea water, and also in the enormous saline deposits of Stassfurt. From the latter source the element is obtained by the manufacturer.

In the laboratory we can prepare bromine from sodium or potassium bromide, by mixing either salt with manganese dioxide and sulphuric acid (exactly as in the preparation of chlorine from sodium chloride). The sulphuric acid liberates hydrobromic acid, which, in the presence of the manganese dioxide, is deprived of its hydrogen, leaving the element free.

The operation is carried out in a retort, and the bromine collected in a cooled receiver.

Bromine dissolves in water, the solution is called bromine water and has the red colour of bromine. Like chlorine water, this solution of bromine has bleaching powers, but less strong than those of chlorine. Bromine also combines readily with metals.

Chlorine can turn bromine out of its compounds. If we make a solution of potassium bromide and add chlorine water to it, the chlorine seizes the potassium and forms potassium chloride, and the bromine is set free. The solution, therefore, turns reddish with the liberated bromine.

Iodine is a steel grey shining crystalline solid, with a lustre like a metal. When gently heated, it melts, and quickly passes into a vapour having a magnificent violet colour. In combination, iodine is present in small quantities in sea water, and is taken up by certain sea plants. Formerly the element was entirely obtained from these sea weeds, but now it is chiefly got from Chili saltpetre, which is found to contain a small quantity of sodium iodate mixed with it.

In the laboratory the element can be obtained from potassium iodide, just as bromine is from the bromide.

Both bromine and chlorine can turn iodine out of its compounds ; if, therefore, we add a few drops of either bromine water or chlorine water to a solution of potassium iodide, the iodine is liberated, and the potassium is taken by the bromine or chlorine. We can tell when free iodine is produced by such reactions as these, by a beautiful and delicate test.

When *free* iodine comes in contact with starch, an intensely blue coloured compound is formed, therefore, before adding the bromine or chlorine, we first put into the solution of potassium iodide a little thin starch emulsion (made by pouring boiling water upon a little powdered starch). So long as the iodine is in combination with potassium, it is incapable of uniting with the starch, but the instant the smallest trace of it is set free (by the addition of the bromine or chlorine) the blue colour is formed.

Compounds of the halogens with hydrogen.—Fluorine, bromine, and iodine each unite with hydrogen, forming respectively hydrofluoric acid, hydrobromic acid, and hydriodic acid. These three compounds are colourless fuming gases, strongly resembling hydrochloric acid in their properties.

Fluorine combines with hydrogen with explosion, even in the dark. Chlorine and hydrogen do not unite in the dark, but do so with explosion when exposed to bright light.

Bromine vapour and hydrogen do not combine by the influence

of light, but do so in contact with a flame; while iodine and hydrogen require to be strongly heated in order to cause them to unite. This illustrates the fact that these four elements gradually become less and less chemically active, as we pass from fluorine to iodine.

Hydrofluoric acid has one remarkable property not possessed by either of the others, namely, its power of attacking glass. On this account it is used to etch glass. The glass is first coated with wax, and the design to be etched is then scratched through the wax. The whole is then exposed either to gaseous hydrofluoric acid or to a solution of the gas in water. The acid eats into the glass where the wax has been removed, and in this way etches the surface. Owing to its action on glass, the acid has to be preserved in either leaden or guttapercha bottles.

CHAPTER XIV

WE have learnt (p. 45) that the hydrogen present in sulphuric acid is expelled or displaced by certain metals, such as zinc, magnesium, or iron. Now the question arises, does it make any difference to the amount of hydrogen obtained, which of these metals we use? In other words, will the same weight of each metal displace the same quantity of hydrogen? This is an important question, and we must try to answer it for ourselves by experiment.

FIG. 76.

Experiment 120.—**Hydrogen displaced by zinc.** Fit up a test-tube as shown in Fig. 76, with a thistle funnel and delivery tube. [The cork and tubes must fit quite tight.] Now weigh out from one to two grams of zinc-foil and place it in the test-tube with a little water. Support the apparatus so that its delivery tube is beneath a cylinder standing in the water trough, and then slowly pour a few drops of strong sulphuric acid down the funnel. Hydrogen is almost immediately set free, and collects in the cylinder. Allow the experiment to go on until the whole of the zinc has dissolved. If necessary one or two more drops of acid can be added.

When the action is finished, the little apparatus is of course filled with hydrogen, but this is compensated for by the air with which it was filled at the beginning having been collected in the cylinder.

The volume of gas collected, is measured as described in Exp. 100. As hydrogen is such a light substance, we may omit any correction for pressure, as the difference this would make will be too slight to effect our result. The temperature must be observed as before.

EXAMPLE.—Weight of zinc used = 1·95 grams.

> Volume of hydrogen as collected = 706 cc.
> Temperature = 15°

(1) 706 cc. of hydrogen at 15°, what volume will it measure at 0°?

$$\frac{706 \times 273}{(273 + 15)} = 670 \text{ cc.}$$

(2) What is the weight of 670 cc. of hydrogen?

> 1 litre of hydrogen (1000 cc.) weighs 0·0896 grams
> therefore 1000 : 670 :: 0·0896 : x
> x = 0·06 grams of hydrogen.

Hence, 0·06 grams of hydrogen are displaced by 1·95 grains of zinc; or, 1 part by weight of hydrogen is displaced by 32·5 parts by weight of zinc

Let us now repeat this experiment, using the metal magnesium instead of zinc. We can use the same apparatus and conduct the operation in the same way, except that the acid must be more dilute, as the action of magnesium is more energetic than that of zinc. We may, however, perform the operation in this case in a still more simple manner.

Experiment 121.—**Hydrogen displaced by magnesium.** Fill a glass cylinder with very dilute sulphuric acid (1 of acid to from 20

to 30 of water) and invert it in a glass dish containing the same liquid (a metal pneumatic trough should not be used). Now weigh out about ½ gram of clean magnesium ribbon, fold it up and carefully introduce it under the mouth of the cylinder. Although magnesium is heavier than water, and therefore sinks in that liquid, it will in this case be buoyed up to the top of the dilute acid by the bubbles of gas which are evolved from its surface. In a very short time the metal will have entirely dissolved, when the gas may be measured in the way already explained.

EXAMPLE.—Weight of magnesium used = 0·48 grams

Volume of hydrogen measured = 466 cc.
Temperature 14° C.

(1) 466 cc. of hydrogen at 14° C. What volume will it be at 0°?

$$\frac{466 \times 273}{(273 + 14)} = 443 \text{ cc.}$$

(2) What is the weight of 443 cc. of hydrogen?

1000 cc. hydrogen weigh 0·0896 grams
therefore, 1000 : 443 :: 0·0896 : x = 0·0397 grams ;

hence 0·0397 grams of hydrogen are displaced by 0·48 grams of magnesium ;

or, 1 part by weight of hydrogen is displaced by 12 parts of magnesium.

It is evident from these experiments that the two metals, zinc and magnesium, differ very widely in their power to displace hydrogen from sulphuric acid, for 12 parts of the latter are *equivalent* to 32·5 parts of the former in this respect. Let us test this matter a little further, using another metal, and a different hydrogen compound.

Experiment 122.—**Hydrogen displaced by sodium.** Take one of the little lead tubes described on p. 42, and counterpoise it on the balance. Then fill the tube with sodium, as there explained, using a clean piece and not exposing it to the air longer than is absolutely necessary. Then weigh the tube and contents. Now take the tube between the thumb and forefinger and quickly introduce it beneath the mouth of a cylinder of water in the pneumatic trough, holding it there until no more gas is evolved. When the gas stops, gently shake the little tube by tapping it against the bottom of the trough so as to be quite sure that the sodium is entirely gone. Measure the gas, and take its temperature as before.

EXAMPLE.—Weight of sodium used = 0·34 grams.

Volume of hydrogen as collected = 173 cc.

Temperature = 12° C.

(1) 173 cc. at 12°, what volume at 0° ?

$$\frac{173 \times 273}{273 + 12} = 165 \text{ cc.}$$

(2) What is the weight of 165 cc. of hydrogen ? ∨

1000 cc. hydrogen weigh 0·0896 grams ;

∴ 1000 : 165 :: 0·0896 : x = 0·0147 grams.

Hence 0·0147 grams of hydrogen are displaced by 0·34 grams of sodium ;

or, 1 part by weight of hydrogen is displaced by 23 parts of sodium.

This experiment teaches us that sodium is different from both zinc and magnesium in its power of turning hydrogen out of its compounds, for we find that in this respect 23 parts of sodium are equivalent to 12 parts of magnesium, and to 32·5 parts of zinc.

In other words, we may say that 23 parts of sodium, 12 parts of magnesium, and 32·5 parts by weight of zinc, are each *chemically equivalent* to 1 part by weight of hydrogen.

Every element, however, is not capable of displacing hydrogen from its compounds in this simple way, or it would be an easy matter to find the particular weight of each element which was equivalent to 1 part of hydrogen.

In the case, therefore, of a great many of the elements, we have to go to work in a more roundabout manner in order to find what weight of them is equivalent to 1 part by weight of hydrogen. For instance, having discovered the weight of zinc which is capable of displacing one part by weight of hydrogen, we can find the weight of certain other metals which can be displaced by this equivalent weight of zinc.

Experiment 123.—Add a few drops of a solution of copper sulphate to half a test-tube full of water, and put into the solution a narrow strip of sheet zinc. Notice that the zinc at once begins to blacken, and that in a very short time the blue colour of the solution begins to fade away, and presently to disappear altogether.

Experiment 124.—Pour two or three drops of silver nitrate solution upon a piece of flat glass, and lay a minute piece of zinc upon the liquid, and let it remain still. Look at it with a pocket lens, and notice bright shining crystals of silver beginning to be deposited all round the piece of zinc, and gradually spreading right across the drops of liquid.

These two experiments show us that both copper and silver are displaced from solutions of their salts by the metal zinc, for the black deposit in the first case is simply copper in the form of a very fine powder. We might repeat these experiments, employing magnesium in the place of the zinc, with similar results. In this case, however, the action is slower, and the displacement of the metals is less complete.

In order to ascertain the weight of, say, silver which is displaced by 32·5 parts by weight of zinc, we can proceed as follows.

Experiment 125.—**Silver displaced by zinc.** Weigh out about half a gram of clean zinc foil, place it in a small beaker and pour a solution of silver nitrate upon it, and allow it to stand. As the silver is gradually deposited, the zinc is as gradually dissolved, until finally it has entirely disappeared. By gently feeling with a glass rod it is easy to ascertain when the whole of the zinc is dissolved, because the precipitated silver is quite soft and spongy. [The solution of silver nitrate should not be too dilute, or it may be necessary to decant off the first quantity and add more, in order to provide enough silver for the weight of zinc employed.] When the zinc is completely gone, the liquid must be carefully decanted off, and the precipitated silver washed once or twice by half filling the beaker with water, stirring gently, allowing it to settle and again decanting the liquid. The silver is next transferred to a clean porcelain crucible, which, with its lid, has been counter-poised carefully. This is done by collecting it together as much as possible, and withdrawing it by means of a glass rod, in the manner shown in Fig. 77, the last particles being rinsed out with water. When the whole has in this way been got into the crucible, the water is decanted and the silver drained as dry as possible. The crucible, partly covered with its lid, is then placed in a hot-air oven, heated to about 110° to 120°, in order to dry the silver perfectly ; after which it is allowed to cool, and weighed. It should then be returned to the oven for a second heating, and once more weighed. If this last weighing agrees with the one before, it shows that the substance was perfectly dry.

EXAMPLE.—Weight of zinc used = 0·48 gram.
        ,,   silver obtained = 1·60 grams.
    Then as 0·48 : 32·5 :: 1·6 : x.

$$x = \frac{32·5 \times 1·6}{0·48} = 108·3 = \text{Weight of silver displaced by 32·5 parts}$$

by weight of zinc.

By this experiment, then, 108·3 parts by weight of silver are equivalent to 32·5 parts by weight of zinc. But this weight of zinc is equivalent to 1 part by weight of hydrogen;

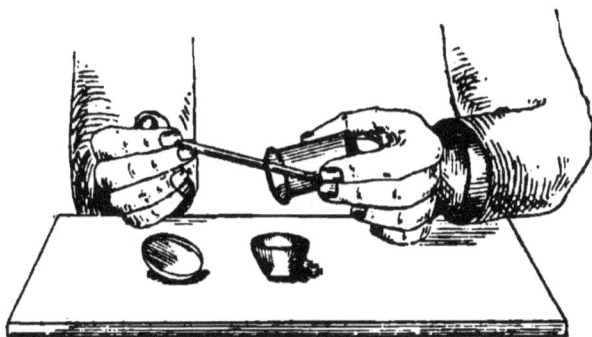

FIG. 77.

therefore we say that 108·3 parts by weight of silver are equivalent to 1 part by weight of hydrogen.

By a similar experiment, using a solution of copper sulphate instead of silver nitrate, we could ascertain the weight of copper that is equivalent to 1 part of hydrogen.

When we examine the compound of hydrogen with chlorine, we find that the elements combine together in the proportion of one part *by weight* of hydrogen with 35·5 parts by weight of chlorine. We say, therefore, that 35·5 parts of chlorine are equivalent to 1 part of hydrogen.

By Exp. 122 we learnt that 23 parts of sodium are equivalent to 1 part of hydrogen. Hence we have—

35·5 parts by weight of chlorine equivalent to 1 part of hydrogen.
23   ,,   ,,   sodium   ,,   1   ,,

Now we know that sodium and chlorine themselves

combine together, and the question arises, do they unite in the proportion of these equivalent numbers?

When the compound sodium chloride is analyzed, it is found to contain the two elements in the proportion 1 part of sodium to 1·535 parts of chlorine,

$$\text{but } 1 : 1\text{·}535 :: 23 : 35\text{·}5;$$

therefore the proportion in which sodium and chlorine separately combine with hydrogen, is the same as that in which they unite together. We may say, therefore, that 23 parts of sodium are equivalent to 35·5 parts of chlorine.

Again, from Exp. 97 it was found that—

1 part of hydrogen combines with 8 parts of oxygen by weight,

and from Exp. 94 that—

12 parts of magnesium combine with 8 parts of oxygen by weight.[1]

But we have also learnt that 12 parts of magnesium and 1 part of hydrogen are chemically equivalent, therefore they are each equivalent to 8 parts of oxygen.

Similarly since 23 parts of sodium and 8 parts of oxygen are each equivalent to 1 part of hydrogen, they are equivalent to each other, and if we were to examine the compound they form when they unite, we should find that it contained sodium and oxygen in the proportion of 23 to 8.

And again, since 35·5 parts of chlorine and 12 parts of magnesium are each equivalent to 1 part of hydrogen, they are equivalent to each other; and when the compound of magnesium and chlorine is analyzed it is actually found to contain these elements in the proportion of 12 parts of magnesium to 35·5 of chlorine.

By Exp. 96 we have learnt that sulphur and oxygen combine in equal weights, that is—

8 parts of sulphur combine with 8 parts of oxygen;

but

1 part of hydrogen combines with 8 parts of oxygen.

[1] The proportion 1·5 magnesium to 1 oxygen, obtained by Exp. 94, is the same as 12 to 8.

Do sulphur and hydrogen follow the same rule as in the former cases, and combine in the proportion of 8 to 1? When the compound of sulphur and hydrogen is examined, it is found that it contains the elements in the proportion, sulphur 16 parts to hydrogen 1 part.   Instead of 8 to 1, the proportion, therefore, is 8 × 2 of sulphur to 1 part of hydrogen. When chemists had investigated a large number of cases where elements which combined together, also combined separately with some other element, they found that in some instances, the proportion in which the elements *separately* combined with one common element, was the *same* as that in which they united together; while in other cases (as in the above example) it is a *simple multiple* of that number. These facts are expressed in the third general law of chemical combination, known as the

Law of reciprocal proportions, or the law of *equivalent proportions. The weights of different elements which separately combine with a constant weight of another element, are either the same as, or are simpler multiples of the weights of these elements which combine with each other.*

These various weights of the elements are called their *combining proportions*, or the *equivalent weights* of the elements. They represent the proportions by weight in which the elements are able to combine together, relative to 1 part by weight of hydrogen.

Experiments on Neutralizing Acids.

Experiment 126.—Counterpoise a small beaker, and then weigh into it 100 grams of strong oil of vitriol. Pour this into a litre flask, rinsing the beaker three or four times with water, and carefully pouring each rinsing into the flask, so that the whole of the acid may be transferred without loss. Now fill the flask up to the mark with distilled water. This litre, therefore, contains 100 grams of sulphuric acid;[1] and as a litre is 1000 cc., each cc. of the liquid will contain 0·1 gram of acid. Transfer this solution to a clean stoppered bottle. Next weigh out 25 grams of solid caustic soda. [This weighing must be done as quickly as possible,

[1] In reality this is not quite exact, as even the strong sulphuric acid contains a *small* quantity of water; but it will be near enough for our present purpose.

as caustic soda absorbs moisture from the air very rapidly.] Dissolve this in water in a half-litre flask, and then fill it to the mark with more water. Since the half litre (that is, 500 cc.) contains 25 grams of caustic soda, 1 cc. will therefore contain 0·05 gram.

By means of a pipette (Fig. 65), transfer 50 cc. of this solution to a beaker, and add to it one or two drops of litmus solution.

Fill a burette (Fig. 64) with the prepared sulphuric acid, and allow it to run, in small quantities at a time, into the beaker, which should be stood on a white tile, or piece of white paper. As the point is approached at which the alkali is neutralized, each drop of acid produces a temporary reddening of the liquid as it falls in. The liquid should be stirred, and the addition of acid stopped the instant the blue litmus is permanently reddened. Read and note how many cc. of acid have been added in order to so neutralize the alkali, and then repeat the experiment so as to obtain two results which agree.

Example.—50 cc. of the caustic soda required 30 cc. of the sulphuric acid. Since 1 cc. of the acid contains 0·1 gram of sulphuric acid, 30 cc. will contain 3 grams, therefore 3 grams of sulphuric acid are neutralized by 50 cc. of the caustic soda, or 2·5 grams ; or 100 grams of acid are neutralized by 83·3 grams of caustic soda.

Experiment 127.—Weigh out 25 grams of caustic potash, and dissolve it in a half-litre flask, and fill to the mark with water. Take 50 cc. of this solution and proceed exactly as in the last experiment. Note the volume of acid required to exactly neutralize the alkali.

Example.—50 cc. of caustic potash solution required 22 cc. of acid, or 2·2 grams. Therefore 2·2 grams of sulphuric acid are neutralized by 2·5 grams of caustic potash ; or 100 grams of acid are neutralized by 113·6 grams of caustic potash.

According to these experiments, therefore, 83·3 parts by weight of caustic soda are equivalent, in their power of neutralizing sulphuric acid, to 113·6 parts of caustic potash.

Let us now see whether these two alkalies stand in the same relation towards another acid.

Experiment 128.—Take 50 cc. of ordinary concentrated hydrochloric acid, by means of a pipette, and dilute it to half a litre. (For our present purpose it is not necessary to know the weight of the acid we are going to use.)

Now withdraw 50 cc. of this diluted acid with a pipette, transfer it to a small beaker and colour it with litmus.

Fill a burette with the caustic soda solution prepared in Exp. 126, and add it gradually to the acid until the solution is exactly neutral.

Now repeat this with the caustic potash solution, taking another 50 cc. of the acid and neutralizing it with the potash.

Calculate as in the former experiment the different weights of these two alkalies required to neutralize equal quantities of hydrochloric acid ; and see whether they are in the same ratio as 83·3 to 113·6.

CHAPTER XV.

THE ATOMIC THEORY.

THE three laws of chemical combination, namely, the *law of constant proportion*, the *law of multiple proportion*, and the *law of equivalent proportions*, are the general expressions of a vast number of well-established *facts*. By experiments such as those described in the foregoing chapters, chemists have discovered the *facts* that chemical combination between the elements takes place according to these rules or laws, facts which must be regarded as indisputable.

The human mind, however, is not satisfied with facts; it always asks the question, Why? In order to account for, and to explain these extraordinary *facts*, in order to give some satisfactory answer to the question, *Why do the elements unite in definite and in multiple proportions?* chemists adopt a *theory*. A *law* is the general expression of discovered *facts ;* a *theory* is a hypothesis or guess which we make in order to explain the facts. If the hypothesis fits all the facts, it is then called a *theory;* and so long as no new facts are discovered which do not square with the theory, it is accepted as an explanation of the facts ; but if fresh facts are found out which will not harmonize with the theory, then the theory must be given up, and a new one proposed.

The theory which chemists have adopted to explain the facts connected with chemical combination, is known as the *atomic theory*, or sometimes *Dalton's atomic theory*, from the name of the chemist who first proposed it.

Dalton's theory consists of four hypotheses or surmises :—

(1) That all the elements, whether solid, liquid, or gaseous, consist of a vast number of minute indivisible particles or *atoms*.

(2) That the atoms of any one element have all the same weight.

(3) That the atoms of different elements have different weights, and that these weights stand in the same ratio to one another as the numbers expressing the combining proportions of the elements.

(4) That chemical combination consists simply in the union together of *atoms;* the atoms being held together by the operation of the force, chemical affinity.

Let us see now how this *theory* will help us to understand the known *facts* about the manner in which elements combine together ; and first as to the law of constant proportion. When sodium combines with chlorine, we have seen (p. 142) that the proportion in which these two elements unite, is sodium 23 parts and chlorine 35·5 parts. Analysis shows that this proportion is constant, that the compound *sodium chloride* invariably contains the elements, sodium and chlorine, in these proportions.

According to the atomic theory, sodium chloride is the product of the union of *atoms* of sodium with atoms of chlorine, and the relative weights of these atoms is expressed by the combining proportions of the elements, namely, 23 and 35·5. Since, by the theory, atoms are *indivisible*, it follows that the compound produced by the union of one atom of each of these two elements must always have the same composition. The proportion cannot be 22 : 35·5, or 23 : 30, because neither of these represents the *relative weights of the indivisible atoms* of sodium and chlorine.

Again, as to the law of multiple proportions. We saw (p. 118) that carbon unites with oxygen in two proportions, producing two compounds.

In the first of them the elements are present in the proportion, carbon 1 part, and oxygen 1·33 parts. From the figures on p. 142, it will be seen that the combining proportion for oxygen is 8. But the weight of carbon which bears the same ratio to 8 as 1 bears to 1·334 is 6.

$$1 : 1·334 :: 6 : 8$$

Therefore carbon and oxygen are contained in this compound in the proportion of carbon 6 parts and oxygen 8 parts. The composition of this compound, like that of the sodium chloride, is invariable; one indivisible atom of carbon unites with one indivisible atom of oxygen, the relative weights of these two atoms being as 6 : 8.

In the second compound of carbon and oxygen, the carbon and oxygen are present in the proportion, carbon 6 parts and oxygen 16 parts. The atomic theory accounts for this. The one atom of carbon is here combined with *two* atoms of oxygen, each having the relative weight of 8. If carbon is to combine with more than 1 atom of oxygen, since atoms are indivisible, it must unite with at least 2 atoms.

In the five compounds of nitrogen and oxygen (p. 118) the two elements are present in the proportions—

(1) Nitrogen : oxygen = 1 : 0·57 or 14 : 8
(2) Nitrogen : oxygen = 1 : 1·14 or 14 : 16
(3) Nitrogen : oxygen = 1 : 1·71 or 14 : 24
(4) Nitrogen : oxygen = 1 : 2·28 or 14 : 32
(5) Nitrogen : oxygen = 1 : 2·85 or 14 : 40

The increase in oxygen takes place regularly by the addition of a weight of that element equal to 8, that is, by a weight equal to the *combining proportion* of oxygen, which according to Dalton's hypothesis is the relative weight of the atom of oxygen.

Atomic Weights of the Elements.—As an outcome of the atomic theory, the numbers which formerly represented the combining proportions of the elements became invested with a new significance, and were called the *atomic weights* of the elements. Since Dalton's day, however, with the growth of chemical knowledge, it has been found necessary to change many of the numbers which were by him regarded as the atomic weights; and in many cases, instead of the number which expresses the *combining proportion*, it is some *multiple* of this number which is to-day accepted as the true atomic weight. For example, the combining proportion of oxygen is 8, its atomic weight is now regarded as 16. The combining proportions of magnesium and of sulphur are respectively 12 and 16, but we now assign the numbers 24 and 32 as the atomic weights of these elements.[1]

The numbers which are at the present time accepted by chemists as the approximate atomic weights of the elements, are given in the third column of the table on the cover. The student should make himself familiar with the atomic weights of a number of the commoner elements. Referring to the table, we there see that the atomic weight of carbon is 12; let him remember that this simply means, that the smallest weight of carbon which enters into chemical combination is 12 times as heavy as the smallest weight of hydrogen which is capable of uniting with another element; or, in the words of the atomic theory, that the carbon atom is 12 times as heavy as the hydrogen atom. The atomic weight of oxygen and of nitrogen are 16 and 14, that is to say, the atoms of these elements are respectively 16 and 14 times heavier than the hydrogen atom. Hydrogen being the lightest known substance, the weight of its atom is taken as the unit, and in the table therefore its atomic weight is given as 1.

Atoms and Molecules.—Chemists believe that, generally speaking, atoms are unable to exist alone, that is, in a free or uncombined state; but that the moment an atom is expelled from combination, it immediately combines with some other atom or atoms; either with an atom like itself or an atom of some other element. The smallest particle of matter, therefore, which is capable of independent existence, consists usually of a little group or system of atoms, bound together by chemical affinity. These groups of atoms are called *molecules*. If the atoms which are thus associated together to form a molecule are atoms of different elements, the molecule is the *molecule of a compound*, but when an atom is combined only with other atoms of the same element, the molecule is a *molecule of an element*. For example, a molecule of the

---

[1] There are many considerations which now influence chemists, in deciding upon the particular number which shall be assigned as the atomic weight of an element, but the study of these matters the student may leave until a later stage, and he will find them fully set forth in more advanced text-books. See Newth's "Inorganic Chemistry."

compound sodium chloride consists of one atom of sodium united to one atom of chlorine, while a molecule of the element chlorine is composed of an atom of chlorine united to another chlorine atom.   To the mind of the chemist all kinds of matter, say a piece of chalk, or a drop of water, present the appearance of an innumerable multitude of these tiny molecules[1] or groups of atoms, packed more or less closely together.   When a solid substance, such as ice, is melted, the mental picture which the chemist sees of the operation, is that the molecules are thrust further apart from each other, and again when the liquid water is converted into steam, he sees the molecules separated to a still greater distance from each other.   But in all these changes, he sees that the atoms which compose the molecules still remain associated together; each water molecule is a little group of three atoms (two of hydrogen and one of oxygen), and these three atoms, held together by chemical affinity, remain closely bound together whether the water be solidified to ice or vaporized to steam.   These changes only effect the molecule *as a whole*, they are, therefore, only physical changes; but as soon as we disturb the *composition* of the molecule, as soon as we break through the force which unites the atoms together in the group, then we produce a chemical change upon the substance.   We then separate the constituent atoms in the water molecules, and obtain two substances, namely, hydrogen and oxygen, which are both entirely different in properties from the original water.

We may see some analogy to the relations between atoms and molecules, in many of the systems of the heavenly bodies. Thus, the planet Jupiter with its five moons may be compared to a molecule consisting of six atoms.   The planet and its satellites remain constantly associated together, performing certain movements relative to each other, while, at the same time, the whole system in its undisturbed unity courses through space on its own independent orbit.   This illustration, however, is incomplete in this point; we know that the force,

---

[1] The word "molecule" means little mass.   They are so extremely minute that a single drop of water is made up of countless millions.   It is utterly beyond the power of the strongest microscope to detect them.

which binds the planet and its moons together and keeps them united in a single system, is the same as that which regulates the movements of the united system as it travels round the central sun, namely, *gravitation;* but the force which binds the atoms together in a molecule, is the force we call *chemical affinity*, and it appears to be a totally different force from any of the merely physical forces which operate between the molecules themselves.

An atom may be defined as *the smallest weight of matter which can take part in a chemical change:* while we define a molecule as *the smallest weight of matter which can exist in the free state.*

CHAPTER XVI

Chemical Symbols.—In the second column of the table on the cover, the symbols are given which chemists use to denote the different elements. In a number of cases these symbols are merely the first letter of the name, such as Carbon, C; Hydrogen, H; Oxygen, O; Sulphur, S; and so on. In others it is either the first or the second letters, as Bromine, Br; Silicon, Si; or, the first and one other that is prominent in the name, such as Chlorine, Cl; Manganese, Mn. In some instances the symbol is taken from the Latin name, thus— Copper (*cuprum*), Cu; Iron (*ferrum*), Fe; Silver (*argentum*), Ag.[1] These symbols are not intended to be used as mere shorthand signs only, but they stand in all cases for *one atom* of the various elements. Thus, the symbol H stands for *one atom of hydrogen;* the symbols O, K, Na, Cl, represent *one atom* of oxygen, potassium, sodium, and chlorine respectively, and not any indefinite quantity of these substances. They are, in a strict sense, *atomic symbols*, that is, symbols of the atoms, and they should therefore never be employed by the student as mere abbreviations for the *names* of the elements.[2]

If we know how many atoms of an element go to make up

---

[1] The student must make himself familiar with the symbols of all the more common elements, such as those contained in the list on p. 8.

[2] It is true that chemists often fall into the habit of employing these symbols as mere abbreviations in writing, but the habit is to be deprecated, especially in the beginner, as its practice tends greatly to obscure the true significance of the symbols ; namely, that of representing a definite quantity (one atom) of the various elements.

the molecule of that element, we can express that knowledge by means of these symbols, by the use of numerals placed immediately after them. Thus, in the case of the elements hydrogen, oxygen, nitrogen, chlorine, and some others, the molecules consist of groups of two atoms, and as the symbol stands for one atom, the molecules of these elements are represented by H_2, O_2, N_2, Cl_2 respectively. The symbol H_2, therefore, means *one molecule of hydrogen.*

Chemical Formulæ.—By means of these symbols, the composition of compounds is represented. This is accomplished by placing the symbols for the various elements in the compound, side by side. Thus, the compound obtained when sodium (symbol Na) and chlorine (symbol Cl) combine together, is expressed by the united symbols of the two elements, NaCl. Such an arrangement of symbols standing for a compound is called the *formula* for that compound. Such a formula, however, expresses the *composition of a molecule* of the compound; and as Na stands for *one atom* of sodium, and Cl for *one atom* of chlorine, the molecule of sodium chloride (as represented by this formula) will consist of one atom of each of the two constituents. HCl, in like manner, is the *molecular formula* for the compound of hydrogen and chlorine, and the formula states that the molecule of hydrochloric acid consists of one atom of hydrogen united to one atom of chlorine.

When the molecules of any compound contain more than one atom of any particular element, the fact is indicated by the use of small numbers placed immediately after the symbol of that element; thus H_2O is the molecular formula for water, a compound containing in its molecule two atoms of hydrogen and one of oxygen. NH_4Cl is the formula for one molecule of ammonium chloride, in which the molecule is composed of one atom of nitrogen, four atoms of hydrogen, and one of chlorine. Very often in formulæ a little more complex, it is necessary to indicate the presence of certain groupings of atoms. For this purpose brackets are employed, thus $(NH_4)_2SO_4$ represents a molecule containing four atoms of oxygen, one atom of sulphur, eight atoms of hydrogen, and two of nitrogen, the last two elements being associated together in two groups, each group consisting of one atom of nitrogen and four of hydrogen.

When it is necessary to indicate more than one molecule of any substance, large numerals are placed before the symbol or formula, thus $2H_2$ means two molecules of hydrogen; $5H_2O$ signifies five molecules of water.

Chemical Equations.—One of the chief uses of these symbols and formulæ is to enable chemists to express, in a condensed and precise form, a considerable amount of information respecting the various chemical changes which it is their business to study. These changes are called *chemical reactions*, and they are expressed in the form of *equations*. The symbols and formulæ of all the materials undergoing change are placed to the left, and those of the new products resulting from the action, on the right.

Thus, when mercury and iodine unite to form mercuric iodide, the change would be expressed by the following equation—

$$Hg + I_2 = HgI_2.$$

This means that one atom of mercury combines with two atoms (one molecule) of iodine, and forms one molecule of mercuric iodide, which contains one atom of mercury and two atoms of iodine.

By means of similar equations we can express all chemical changes which are understood. Take, for example, the various reactions by which the element hydrogen was obtained, Chapter VI.

The action of sodium on water is represented thus—

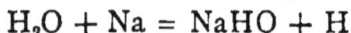

$$H_2O + Na = NaHO + H$$

which means that when one molecule of water is acted on by one atom of sodium, one atom of hydrogen is displaced, and a molecule of sodium hydroxide is formed. Or, again, the equation

$$H_2SO_4 + Zn = ZnSO_4 + H_2$$

means that one atom of zinc acts on one molecule of sulphuric acid, displacing two atoms (one molecule) of hydrogen and forming a molecule of zinc sulphate.

As matter can neither be created nor annihilated, every single atom which figures on the left side of an equation must be accounted for on the other side. In no chemical change is there such a thing as an atom becoming lost or destroyed, they are only made to enter into fresh combinations. Before we can correctly write down any equation, therefore, we must *know* what these new combinations are. For example, when sodium chloride, sulphuric acid, and manganese dioxide are mixed together, a chemical change takes place,

$$2NaCl + 2H_2SO_4 + MnO_2 =$$

If we do not *know* what compounds are actually produced, we cannot complete the equation. We must not simply *re-arrange the symbols*, even although we only make use of the same number of the identical symbols which are here given. There might be fifty possible ways of arranging the symbols, but there is only one which represents what actually happens when chemical action takes place between these compounds.

The three Modes of Chemical Action.—It was stated on p. 14 that chemical action takes place according to three general modes; these are—

(1) *By the direct union of two molecules to form a more complex molecule.*

Experiment 129.—Fill a glass cylinder with gaseous hydrochloric acid (as described in Exp. 104), and cover it with a glass plate. Take another similar cylinder and pour into it a few drops of the strongest ammonia; allow the liquid to run round the glass, and pour out any excess. This vessel will now contain a quantity of ammonia gas. This might be proved by carefully smelling the contents of the cylinder.

Now invert one cylinder upon the other, mouth to mouth, and notice that the two colourless gases combine to form a white solid, which appears as a cloud or smoke.

This solid is ammonium chloride; the molecules of the two gases, hydrochloric acid and ammonia, have united to form the more complex molecules of ammonium chloride. This is expressed by the equation—

$$NH_3 + HCl = NH_4Cl.$$

(2) By the exchange of atoms in different molecules.

Experiment 130.—Dissolve a small quantity of mercuric chloride in water, in a test-tube (this substance dissolves very slowly in cold, but more readily in hot water), and dip into the solution a strip of clean copper foil. Notice that at once the reddish copper becomes coated with a film of mercury, which makes it look as though it had been silvered. The mercury in the solution of the mercuric chloride has been *displaced* by the copper, and an amount of copper *equivalent to this quantity of mercury* has gone into the solution. We express this change as follows—

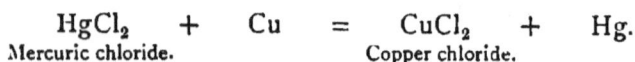

$$HgCl_2 \quad + \quad Cu \quad = \quad CuCl_2 \quad + \quad Hg.$$
Mercuric chloride.   Copper chloride.

If the action is allowed to go on long enough, the *whole* of the mercury will be thrown out or displaced by copper.

We can easily convince ourselves that the solution contains copper chloride by making three little tests.

Experiment 131.—(*a*) Pour a few drops of copper chloride solution into water in a test-tube, and add ammonia. Notice the deep-blue colour produced. This is a characteristic test for copper.

(*b*) Add ammonia to a dilute solution of mercuric chloride. Observe that a white precipitate is formed, but no blue colour.

(*c*) Now add ammonia to the mercuric chloride solution in which the copper has been placed for a short time and then removed. Notice that a white precipitate is produced, and also that the solution turns blue.

Experiment 132.—Pour a few drops of hydrochloric acid solution into water in a test-tube, and add a little silver nitrate solution. A white precipitate at once forms. The silver in the silver nitrate has changed places with the hydrogen of the hydrochloric acid according to the equation—

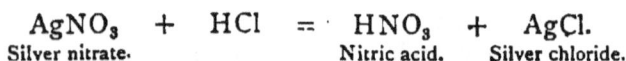

$$AgNO_3 \quad + \quad HCl \quad = \quad HNO_3 \quad + \quad AgCl.$$
Silver nitrate.   Nitric acid.   Silver chloride.

This *mutual* exchange of atoms, which takes place when both the reacting substances are compounds, is often spoken of as **double decomposition.**

(3) By the rearrangement of atoms within the molecule.—The elementary student will not meet with any examples of chemical change which belong to this class.

As a further example of the use of chemical equations, illustrations may be given of the formation of salts by the action of acids upon bases (see p. 69).

(*a*) With metallic oxides—

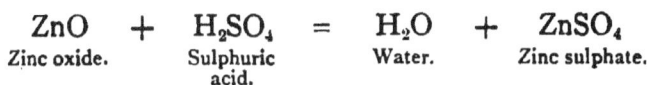

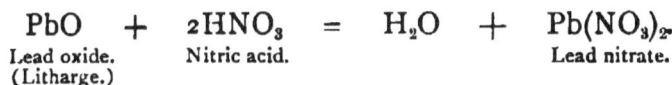

$$ZnO + H_2SO_4 = H_2O + ZnSO_4$$
Zinc oxide.　Sulphuric acid.　Water.　Zinc sulphate.

$$PbO + 2HNO_3 = H_2O + Pb(NO_3)_2$$
Lead oxide. (Litharge.)　Nitric acid.　　Lead nitrate.

(*b*) With metallic hydroxides—

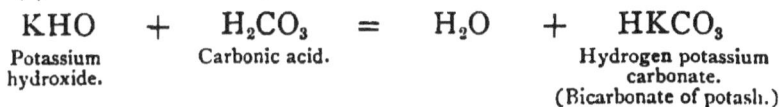

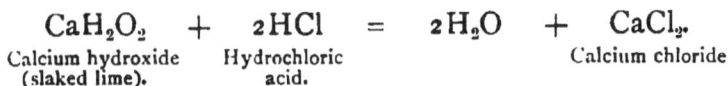

$$KHO + H_2CO_3 = H_2O + HKCO_3$$
Potassium hydroxide.　Carbonic acid.　　Hydrogen potassium carbonate. (Bicarbonate of potash.)

$$CaH_2O_2 + 2HCl = 2H_2O + CaCl_2$$
Calcium hydroxide (slaked lime).　Hydrochloric acid.　　Calcium chloride

(*c*) With ammonia—

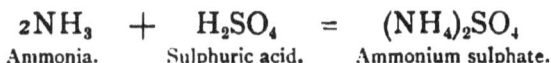

$$2NH_3 + H_2SO_4 = (NH_4)_2SO_4$$
Ammonia.　Sulphuric acid.　Ammonium sulphate.

CHAPTER XVII

THE QUANTITATIVE SIGNIFICANCE OF CHEMICAL EQUATIONS.

WE have seen that a chemical equation tells us what becomes of the various substances which undergo chemical change. By means of these equations, chemists state exactly what are the various substances resulting from such changes. But there is still more information conveyed by these equations, for they actually tell us the *exact quantities* of all the different substances which undergo change, and also of the substances resulting from the action.

We have learnt that a symbol stands for an atom, and also that different atoms have different weights. Thus, the symbol O stands for one atom of oxygen ; but an atom of oxygen is sixteen times as heavy as an atom of hydrogen, therefore the symbol O stands for a quantity of oxygen which is sixteen times as heavy as the quantity of hydrogen represented by the symbol H. We have agreed to take the hydrogen atom as the unit, and call its weight 1 ; H therefore stands for 1 part by weight of hydrogen, and O for 16 parts by weight of hydrogen. In the same way the symbols Na, Mg, S, stand not only for sodium, magnesium, and sulphur, but for 23 parts by weight of sodium, 24 parts by weight of magnesium, and 32 parts by weight of sulphur. This fact will obviously give to a chemical equation a quantitative meaning. Thus, the equation—

$$H_2 + O = H_2O$$

not only means that when hydrogen and oxygen combine, water is produced, but it also signifies that 2 parts by weight of hydrogen unite with 16 parts by weight of oxygen, and yield 18 parts by weight of water—

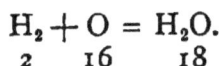

$$H_2 + O = H_2O.$$
$$2 \quad 16 \quad\;\; 18$$

Or, again, the equation—

$$KClO_3 = KCl + 3O,$$

besides meaning that when potassium chlorate is decomposed it yields potassium chloride and oxygen, has also the quantitative significance—

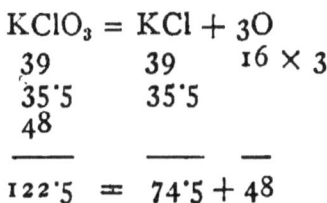

$$KClO_3 = KCl + 3O$$

39	39	16×3
35·5	35·5	
48		
———	———	——
122·5	= 74·5	+ 48

122·5 parts of potassium chlorate yield 74·5 parts of potassium chloride and 48 parts of oxygen, by weight.

Bearing these facts in mind, it is quite easy to calculate how much material we must use to produce any required quantity of any particular product of a chemical change. For instance, suppose we wish to know what weight of oxygen could be produced from, say, 10 grams of potassium chlorate. Then from the above equation we find that 122·5 parts by weight of potassium chlorate give 48 of oxygen, hence we have the *rule of three* sum—

$$\text{As } 122·5 : 10 :: 48 : x$$
Grams of potassium chlorate. Grams of oxygen.

$$x = \frac{48 \times 10}{122·5} = 3·9 \text{ grams of oxygen ;}$$

therefore 10 grams of potassium chlorate will give 3·9 grams of oxygen gas.

Again, suppose we require to find out the weight of magnesium oxide which is produced when four grams of magnesium are burnt, either in air or in pure oxygen. We first write the equation which expresses the chemical change that takes place—

$$Mg + O = MgO$$
$$24 + 16 = 40$$

24 parts by weight of magnesium give 40 parts of magnesium oxide, how much will 4 grams yield?

$$As \underbrace{24 : 4}_{\text{Grams of magnesium.}} :: \underbrace{40 : x}_{\text{Grams of magnesium oxide.}}$$

$$x = \frac{40 \times 4}{24} = 6\cdot6 \text{ grams of magnesium oxide.}$$

Therefore, 4 grams of magnesium on burning, will form 6·6 grams of magnesium oxide.

Let us take one more example. We wish to make 10 grams of hydrochloric acid gas, What weight of sodium chloride and of sulphuric acid shall we require? As before, we write out the equation for the preparation of hydrochloric acid—

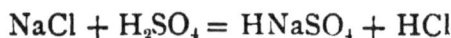

$$NaCl + H_2SO_4 = HNaSO_4 + HCl$$

	23	2	1	1
	35·5	32	23	35·5
		64	32	
			64	

$$58\cdot5 + 98 = 120 + 36\cdot5$$

36·5 parts of hydrochloric acid require for their production 58·5 parts of salt and 98 parts of sulphuric acid, what will 10 grams require?

$$As \underbrace{36\ 5 : 10}_{\text{Grams of hydrochloric acid.}} :: \underbrace{58 : x}_{\text{Grams of salt.}}, \text{ and}$$

$$as \underbrace{36\cdot5 : 10 :: 98 : x'}_{\text{Grams of sulphuric acid.}}$$

$$\text{Then } x = \frac{58 \times 10}{36\cdot5} = 15\cdot9 \text{ (nearly) grams of salt.}$$

$$\text{And } x' = \frac{98 \times 10}{36\cdot5} = 27\cdot1 \text{ grams of sulphuric acid.}$$

Therefore, in order to prepare 10 grams of gaseous hydrochloric acid, we require 15·9 grams of salt and 27·1 grams of sulphuric acid

The method by which we are able to calculate the volume which any given weight of any gas will occupy, has been already explained (p. 101) and exemplified (p. 137), and since we are now able to find the *weight* of gas evolved by any chemical process, we can obviously find its *volume* also. For instance, suppose we require to know the volume of oxygen, measured at 0° and 760 mm., which can be got from 10 grams of potassium chlorate; then we proceed as on p. 159, and find first the *weight* of oxygen. This was found to be 3·9 grams. Then we calculate what will be the volume at N.T.P. of 3·9 grams of oxygen.

Since 1 litre of oxygen weighs 0·0896 × 16 = 1·433 grams, we get the proportion—

$$\text{as } \underbrace{1\cdot433 \ : \ 3\cdot9}_{\text{Grams of oxygen.}} \ :: \ \underbrace{1 \ : \ x}_{\text{Litres of oxygen.}}$$

Therefore $x = \dfrac{3\cdot9 \times 1}{1\cdot433} = 2\cdot7$ litres of oxygen at N.T.P.

If, instead of at N.T.P., we wished to know what volume of oxygen measured at say 12° C. and 751 mm. the 10 grams of potassium chlorate would yield, then, after finding first the *weight*, then the *volume at N.T.P.*, we must make the correction for temperature and pressure, thus—

$$\frac{2\cdot7 \times (273 + 12) \times 760}{273 \times 751} = 2\cdot8 \text{ litres at } 12° \text{ C. and } 751 \text{ mm.}$$

Combination of Gases by Volume.—By Exp. 103 it was shown that when hydrogen and chlorine combine, they do so in equal volumes; and also that when they united, there was *no change in the total volume*, for the volume of hydrochloric acid produced, occupied the same space as did the mixed gases before combination. We say, therefore, that one volume of hydrogen combines with one volume of chlorine and forms two volumes of hydrochloric acid.

M

Again, on p. 77, an experiment is described which shows that when hydrogen and oxygen unite, they do so in the proportion two volumes of hydrogen and one volume of oxygen. When these three volumes of mixed gases combine, however, they *do not give three volumes of steam*, but only *two* volumes.

The relation between the volumes of gases which combine, and the volumes of the compounds produced, is found to be equally simple in all cases, and is expressed in the general statement known as the **Law of Gay-Lussac.**

When chemical action takes place between gases, the volume of the gaseous product bears a simple relation to the volumes of the reacting gases.

Avogadro's hypothesis.—In the year 1811, the Italian physicist Avogadro advanced a theory in order to explain the known facts concerning the behaviour of gases. This theory is accepted by all chemists and physicists, and is called Avogadro's hypothesis. It may be thus stated: *Equal volumes of all gases (under the same conditions as to temperature and pressure) contain the same number of molecules.*

If a litre of oxygen contains the same number of molecules as a litre of hydrogen, it will be evident that, if we weigh a litre of each gas, the difference between these weights will also stand for the difference between the weights of a *single molecule* of these two gases. For example, let us suppose, just for the sake of argument, that a litre of oxygen contains a million oxygen molecules, and that on weighing this volume we find it weighs 1·4336 grams. According to Avogadro, a litre of hydrogen would also contain a million molecules; suppose we find that this volume of hydrogen weighs 0·0896 grams. These numbers then express the ratio between the weight of a million oxygen molecules and a million hydrogen molecules. 1·4336 is sixteen times as much as 0·0896,

for 1·4336 : 0·0896 :: 16 : 1

If, therefore, a million oxygen molecules weigh sixteen times as much as a million hydrogen molecules, obviously each oxygen molecule must be sixteen times as heavy as each molecule of hydrogen.

Therefore, in order to find out how much heavier any gaseous molecule is than a molecule of hydrogen, all that is necessary is to compare the weights of equal volumes of the particular gas and hydrogen.

The Densities of Gases.—The density of a gas is the weight of it as compared with the weight of an equal volume of some other gas which is chosen as the standard. Sometimes air is taken as the standard. Then, when we say that the density of chlorine, for example, is 2·45, we mean that a given volume of this gas is 2·45 times as heavy as an equal volume of air. More generally hydrogen is the standard, in which case the density of chlorine is 35·5; that is to say, a given volume of the gas is 35·5 times as heavy as an equal volume of hydrogen.

Air is 14·44 times as heavy as hydrogen; therefore, if we know the density of any gas with reference to one standard, we can readily calculate by "rule of three" what is its density in terms of the other; thus, the density of chlorine is 35·5 compared with hydrogen, what is its density, air = 1?

$$14\text{·}44 : 1 :: 35\text{·}5 : x$$
$$x = \frac{35\text{·}5 \times 1}{14\text{·}44} = 2\text{·}45$$

When we have ascertained the density of a gas as compared with hydrogen, we have also found (according to Avogadro's hypothesis) the ratio between the weight of a molecule of that gas and a molecule of hydrogen.

Molecular Weights of Gases.—The weight of a molecule of any gas, *as compared with the weight of one atom* of hydrogen, is called the molecular weight of that gas, and as the molecule of hydrogen consists of two atoms, therefore the molecular weight of a gas must be double its density.

For instance, the density of chlorine is found by experiment to be 35·5; that is, its molecule is 35·5 times as heavy as a *molecule* of hydrogen; it will, therefore, obviously be seventy-one times as heavy as half a molecule of hydrogen (or one atom of hydrogen). The unit or standard of comparison for densities is the *molecule* of hydrogen, while the unit for molecular weights is the half molecule, or *atom* of hydrogen.

By experimentally finding the densities of gases, and, therefore, learning their molecular weights, chemists are able to ascertain the *atomic weights* of some of the elements; for the smallest weight of an element that is ever found in a volume of gas equal to the volume occupied by one molecule of hydrogen, is taken as the atomic weight of that element.

The Unit Volume.—If Avogadro's hypothesis be true, if equal volumes of all gases contain (under the same conditions of temperature and pressure) the same number of molecules, it follows that the molecules of all gases occupy the same volume; or, in other words, they occupy the same volume as a molecule of hydrogen. Each molecule of oxygen, or chlorine, or hydrochloric acid gas, or water gas (steam), occupies the same space as a hydrogen molecule.

The space occupied by a molecule of hydrogen is called *two unit volumes* (the unit volume being taken as the space occupied by an *atom* of hydrogen, or the atomic volume of hydrogen), hence we can say that a molecule of all gases occupies two unit volumes.

If we bear in mind that gaseous molecules have the same volume, and if we write our chemical equations so as to represent complete molecules, then the equation at once tells us the volumetric relations between the reacting gases and the gaseous products of the change. For instance, the equation—

$$H_2 + Cl_2 = 2HCl$$

not only carries the information that hydrogen and chlorine combine to form hydrochloric acid; and that the proportion by weight is 1 part of hydrogen to 35·5 parts of chlorine, yielding 36·5 parts of hydrochloric acid; but it tells us that 1 molecule (2 unit volumes) of hydrogen, combine with 1 molecule (2 unit volumes) of chlorine, and give 2 molecules (4 unit volumes) of hydrochloric acid gas. We see at once, therefore, from the equation that when these elements combine there is no change of volume; 4 unit volumes of the mixed gases giving 4 unit volumes of the compound.

Again, in the equation—

$$2H_2 + O_2 = 2H_2O$$

we see that 2 molecules (4 unit volumes) of hydrogen combine with 1 molecule (2 unit volumes) of oxygen, and give 2 molecules of water, which, if measured in the *gaseous state*, occupy 4 unit volumes. (These volume relations *only apply to substances in the gaseous condition.*) At a glance, therefore, we see from this equation that when hydrogen and oxygen unite to form steam, there is a contraction in the volume, equal to one-third of the original.

CHAPTER XVIII

In the early days of Chemistry, the word air was used for all gases. Hydrogen was called *inflammable air;* oxygen, *dephlogistigated* air, and so on. At the present time we employ the word only to denote the gas which surrounds and envelopes the earth.

Formerly, air was regarded as one of the four so-called elements—earth, air, fire, and water; now we know that air is in no sense an element, neither is it a compound, but a mixture of several gases, some of which are elements and some compounds.

The exact composition of the atmosphere has been made the subject of close investigation by many chemists, notably Lavoisier, Boyle, Priestley, Cavendish, and Bunsen; so that it has for a long time been supposed that all that there was to know about the composition of the air was thoroughly known. Nevertheless, so recently as 1894 it was discovered that there is a certain gas present in the air which had hitherto escaped observation and been entirely overlooked. This gas has been named *Argon.*

The Two chief constituents of the Air are Oxygen and Nitrogen; all the others are present in comparatively minute quantities. By various experiments we have learnt that when substances burn in the air, they are in reality combining with the oxygen present; we can, therefore, make use of this fact in order to remove this constituent from the atmosphere.

Experiment 133.—Place a small bit of phosphorus, which has been wiped dry with blotting-paper, in a little porcelain dish floating on water in a pneumatic trough. Light the phosphorus and stand a wide-mouthed bottle or cylinder over it. (See Exp. 70, Fig. 51.) As the phosphorus continues burning in the enclosed space, the flame gradually becomes fainter, until finally it goes out. Now notice that the water rises inside the vessel, showing that a part of the air which originally filled it has disappeared. Allow the jar to remain a little while, so that the white fumes of phosphorus pentoxide may have time to clear away, by getting dissolved in the water, and it will be seen that the remaining gas is quite clear and colourless. Of course, if phosphorus would not go on burning in this gas which is left, it is hardly to be expected that a taper would burn in it; but, to make sure, the jar may be removed (by slipping a glass plate beneath its mouth and lifting it out of the trough) and a lighted taper or candle lowered into the gas.

If this experiment be made in a different way, we can roughly measure what volume of oxygen there is in air.

Experiment 134.—Take a long wide glass tube, sealed up at one end, and fitted with a cork at the other. Divide the tube into five equal divisions, which can be marked by slipping indiarubber rings over the tube. Drop a dry piece of phosphorus into the tube and cork it up. Next dip the end into warm water, so as to melt the phosphorus, and make it begin to burn, and then quickly tip the tube up so as to allow the burning phosphorus to run down the inside. In this way it immediately combines with all the oxygen present. After a moment, when the tube has cooled, open the end under water in a trough. Notice that the water rises to the first ring, showing that $\frac{1}{5}$ of the air has been taken out by the phosphorus.

Therefore, $\frac{1}{5}$ of the air consists of oxygen, which is one of the two chief constituents. The remaining $\frac{4}{5}$ consists of the other chief ingredient, namely, *nitrogen*, mixed with the other gases which are present in very small quantities.

The average of a large number of experiments, shows that the proportion of oxygen and nitrogen in the air is very nearly 21 parts of oxygen to 79 of nitrogen, by volume.

The gases which are present in small quantities are water vapour (to a variable extent), argon (about 0·8 per cent,) carbon dioxide (about 0·04 per cent.), and still smaller quantities of ozone, ammonia, and nitric acid.

The Composition of Air by Weight was carefully determined by Dumas in 1841. In order to understand his method, we may make the following experiment :—

Experiment 135.—Heat a quantity of bright copper gauze in a piece of combustion tube, in a gas furnace, and pass a slow stream of air from a gas holder through the tube. Notice that the copper becomes black, and begins to do so first at the end at which the air enters. Collect the gas which passes out of the tube, in the pneumatic trough, and test it by dipping a lighted taper into a jar of it. Note that it behaves exactly as the nitrogen obtained from air in Exp. 133. The copper in this experiment has combined with the oxygen of the air, leaving the nitrogen.

Dumas heated a weighed quantity of copper in a tube, one end of which was connected to a large vacuous globe with a stopcock, previously carefully weighed. On slightly opening the tap, air was slowly drawn over the heated copper. The copper combined with the oxygen, and the nitrogen was received in the globe. In order to remove the other gases which were present in small quantities, the air was first made to pass through a series of U tubes, filled with materials which would absorb the water vapour and the carbon dioxide. Argon was unknown, and, therefore, went in with the nitrogen. The average of a number of experiments, showed that the proportion of oxygen and nitrogen by *weight* was, oxygen 23 parts, and nitrogen 77 parts.

Air a Mixture.—The oxygen and nitrogen in the air are not chemically combined, but only mixed together. The chief proofs of this are the following :—

(1) The composition of the air is not constant, the proportions of oxygen and nitrogen are found to vary slightly. If the air was a chemical compound of these gases, then, according to the law of constant composition, the constituents would always be present in exactly the same proportions.

(2) The proportion of oxygen and nitrogen in the air does not bear any simple relation to the atomic weights of these elements.

(3) The two gases can be separated by mechanical means; such as by dissolving in water (oxygen is more soluble than nitrogen), and by diffusion.

Diffusion of Gases.—The gases in the air have different densities; water vapour = 8, nitrogen = 14, oxygen = 16, carbon dioxide = 22. Why is it that the heavy carbon dioxide does not sink to the ground and form a bottom layer, and the light water vapour rise up above the oxygen and nitrogen? Two chief causes operate to prevent this, and to keep the gases thoroughly mixed. The first is wind and air currents · the second is the property, belonging to all gases, of *diffusion.*

If we take a jar filled with hydrogen and hold it mouth downwards, although the gas is fourteen times lighter than air, it will nevertheless make its escape downwards out of the bottle.

Experiment 136.—Take two soda-water bottles, and fill one with hydrogen and the other with oxygen. Join the bottles by means of a long piece of combustion tube passing through two corks which fit the bottles. Stand the apparatus upright, the oxygen being at the bottom, and leave it for a few hours. The light hydrogen will gradually find its way down the tube into the oxygen bottle, and the oxygen (sixteen times as heavy) will pass up into the hydrogen bottle; so that, after a time, there will be a perfect mixture of the two gases in both bottles. That the gases have thus mixed can be proved by removing the bottles and applying a light to their mouths. If the gas explodes, it shows that the two have mixed.

FIG. 78.

This power that gases have of mixing themselves together is called diffusion. All gases do not move or diffuse at the same rate. We can prove this in the following way.

Experiment 137.—Take a short clay tobacco-pipe, and cement over the mouth of the bowl a piece of cardboard, or better, a thin piece of unglazed earthenware. Attach the stem of the pipe, with a piece of indiarubber pipe, to a U-shaped glass tube, containing some coloured water, as shown in Fig. 78. Now fill a beaker with hydrogen, hold its mouth downwards, and put the tobacco-pipe up into the beaker. Notice that the water in the U-tube is immediately driven down the limb to which the pipe is attached. This is because the light gas hydrogen diffuses through the porous clay pipe much faster than the air inside the pipe can make its way out, and therefore an excess of gas collects inside, and pushes down the water. Now remove the beaker, and notice that the water not only returns to the level, but *rises* considerably in the limb attached to the pipe. The condition of things is now reversed, the hydrogen which had gone inside the pipe is making its way out again faster than air can get in, hence there is a diminution in the volume of gas inside.

The lighter a gas is, the faster does it diffuse.

Graham's Law.—Graham found out the law which regulates gaseous diffusion, which may be thus stated: *The relative rates of diffusion of any two gases are inversely as the square roots of their densities.* For example—

The density of hydrogen is 1; the square root of 1 = 1.

The density of air is 14·44; the square root of 14·44 = 3·8.

Therefore $\dfrac{\text{the rate of diffusion}}{\text{of hydrogen}} : \dfrac{\text{the rate of diffusion}}{\text{of air}} :: 3 \cdot 8 : 1$

in other words, hydrogen diffuses 3·8 times as quickly as air does.

Now since nitrogen is just a little lighter than oxygen, if we pass a mixture of these two gases through a long porous pipe (under suitable experimental conditions) nitrogen will escape through the walls of the pipe a little faster than oxygen, therefore the gas which is delivered out at the end of the pipes will contain rather more oxygen in proportion to nitrogen than that which is sent in.

By driving a stream of *air* through such pipes, it is found in like manner, that what comes out at the end is a little richer in oxygen than the air which is sent in. Therefore the oxygen and nitrogen in air can only have been *mixed* and not combined, or it would not be possible to sift one away from the other by this process of diffusion.

Combustion.—We know that a great many substances will burn in the air. It is customary to call such things *combustibles*, and to speak of the air as a *supporter of combustion.*

We have learnt by numerous experiments that it is because of the oxygen present, that the air supports the combustion of burning bodies ; oxygen, as we have seen, being such a good supporter of combustion. In ordinary language we call a gas *a supporter of combustion*, if it behaves towards common combustibles in the same way that air does ; but in reality there is no distinction between a combustible and a supporter of combustion. For instance, we have seen sulphur burning in oxygen, and we therefore call sulphur a combustible, and oxygen the supporter of combustion. But let us modify the conditions of the experiment.

Experiment 138.—Heat some sulphur in a wide test-tube until it boils and the dark vapour takes fire at the mouth. Then lower into the test-tube a bent glass tube (so bent that it can enter the test tube) through which a gentle stream of oxygen is passing (Fig. 79). As the jet is passed through the flame of sulphur burning at the mouth, the *oxygen is ignited*, and will continue burning in the sulphur vapour when pushed down into the tube.

Under these conditions the oxygen is the combustible, and sulphur vapour is the supporter of combustion. Again, we have seen hydrogen burn in oxygen, but we can reverse the conditions and make oxygen burn in hydrogen.

Experiment 139.—Fill a jar with hydrogen. Hold it mouth downwards, and apply a light to the gas. While the hydrogen is

Fig. 79.

burning, thrust a jet, from which oxygen is slowly issuing, up into the jar (Fig. 80). As the jet passes through the hydrogen flame it is ignited, and goes on burning just as it did in the sulphur vapour.

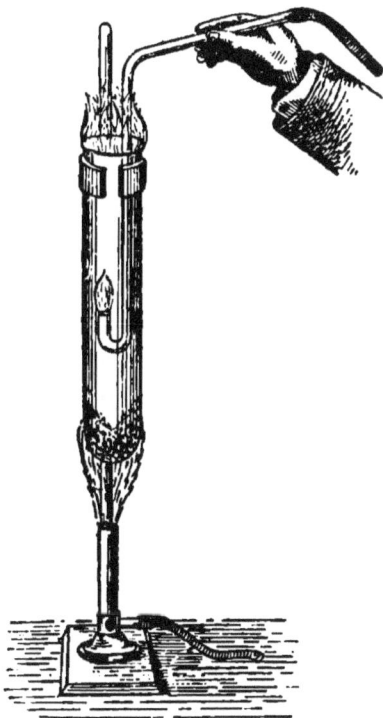

Here oxygen is the combustible, and hydrogen the supporter of combustion.

Combustion is merely the term applied to describe any chemical action which takes place with so much energy as to produce light and heat. All the more common cases of combustion, are the active combination of substances with *oxygen*, that is, they are rapid processes of oxidation; but oxygen is not *necessary* to combustion; we can have cases of combustion in which oxygen does not participate, for we know of many instances in which chemical action takes place between other substances with sufficient energy to give rise to light. For instance—

Experiment 140.—Take a jar of chlorine, and lower into it a jet from which ammonia gas is slowly escaping (the ammonia being obtained by gently warming a strong solution of ammonia, in a little flask; see Exp. 145). Notice that as soon as the jet of ammonia enters the chlorine it at once takes fire, without being lighted, and goes on burning in the chlorine, (Fig. 81). There is here no oxygen present; the ammonia is the combustible, and the chlorine is the supporter of combustion.

We have also seen (Exp. 119) that many metals take fire and burn when brought into chlorine.

Temperature of Combustion.—The actual temperature which is produced by any particular process of combustion depends partly upon how quickly the combustion proceeds. For instance, if we burn some substance in oxygen it burns much faster than when burnt in air, and consequently the temperature is hotter.

FIG. 80.

Experiment 141.—Burn a jet of hydrogen from an oxyhydrogen blowpipe (this is merely a very fine-pointed metal tube with a still smaller one passing down the inside, exactly like an ordinary Herapath blowpipe, only smaller). Hold a little piece of platinum wire in the flame, and note that although the wire becomes very hot, it does not melt. Next hold a little block of hard lime against the flame ; the lime does not seem to get very hot. Now gently turn on the oxygen, so that the hydrogen flame is fed with oxygen. Notice that the flame does not show any more light, but if the platinum wire is held in the flame it is instantly melted, because the flame is now so much hotter ; and if the lime is brought into the flame it at once gets so hot as to emit a very bright light. This is the oxyhydrogen lime-light.

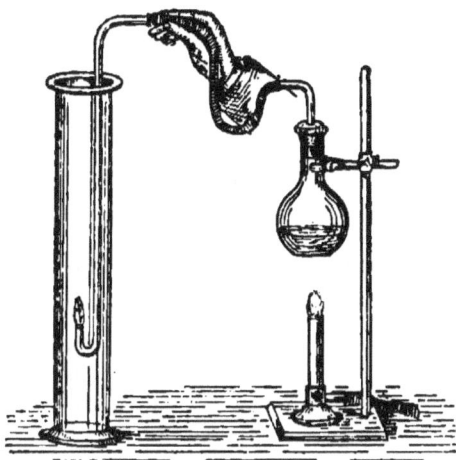

FIG. 81.

In everyday life when we wish to increase the rate of combustion, and consequently raise the temperature of combustion, we increase the draught of air (by the use of bellows, for instance) so that more oxygen is driven against the burning body in a given time.

Ignition point.—The particular temperature at which a substance begins to burn, or to "take fire," is called its ignition point. Sometimes this is lower than the ordinary temperature of the room, in which case the substance takes fire by itself when brought into the air. Obviously such things as these must be kept so that they do not come into contact with the air.

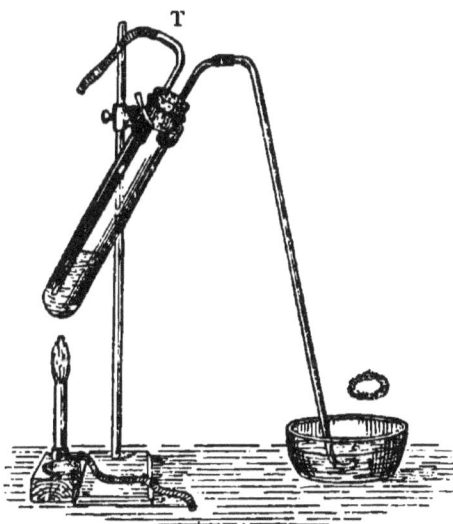

FIG. 82.

Experiment 142.—Place a small quantity of caustic soda solution in a test-tube, and put in it a piece of phosphorus about the size of a pea. Attach a cork with two tubes, arranged as in Fig. 82. First pass a stream of ordinary coal gas through the little apparatus by means of the indiarubber pipe connected with the tube T. This is in order to sweep out the air from the apparatus. Now gently boil the liquid, and a gas is given off called *phosphoretted hydrogen*, which will bubble through the water in the little basin.

In a minute or two, when the coal gas has been expelled, and only the phosphoretted hydrogen is bubbling out, each bubble will take fire as it comes into the air. This gas has a very low igniting point.

In all the familiar processes of combustion, it is necessary to first heat the combustible substance in order to start the combustion; that is to say, the igniting point is *above* the common temperature, and therefore the substance must be heated up to that point before active chemical combination can take place. If after a substance is ignited, it will continue burning by itself, like a candle, or piece of paper, this shows that the *temperature of combustion* is higher than the igniting point; because, as each particle burns, the heat given out is able to set fire to the next particle, and so on.

Heat of combustion is the *amount* of heat as distinguished from the *temperature*, produced by combustion. If we draw a pint of hot water and a gallon of water from the same boiler, the temperature of each sample will be the same; a thermometer placed in each will show the same temperature. But it is obvious that the *amount* of heat in the gallon of water is greater than in the pint; there is, of course, *eight* times as much heat in the one as in the other. From this illustration it will be evident that the *amount* of heat cannot be ascertained by the thermometer. It is really measured by finding out how much water it is capable of heating from 0° to 1°.

That amount of heat which will raise the temperature of 1 gram of water from 0° to 1° is taken as the unit, and is called the *thermal unit* (or calorie).

Now the *amount* of heat produced in any process of combustion, is exactly the same whether the process be slow or quick, therefore it is quite independent of the *temperature* that is produced. For example, the amount of heat given out when a definite quantity of hydrogen is burnt in the air, is exactly the same as that produced when the same quantity of hydrogen is burnt in oxygen, although, as we have seen, the temperature in the latter case is very much higher. Indeed, this is also true if the process of oxidation is so slow that there is no active combustion. Thus, when a quantity of iron slowly rusts, heat is produced; but the process is spread over such a long time, that the iron never even gets warm, the heat being conducted away as fast as it is produced. When the same quantity of iron is burnt in oxygen, the temperature rises enormously, because the process is complete in a few moments, but the actual amount of heat is the same in both cases.

CHAPTER XIX

Nitrogen.—The element nitrogen, as we have seen, is present in the free state in the air; $\frac{4}{5}$ of which is nitrogen. It can be obtained from the air by removing the oxygen either with phosphorus (Exp. 133) or by means of heated copper (Exp. 135).

Nitrogen, combined with other elements, is present in a number of compounds, from some of which the element is readily expelled.

One of the commonest of the compounds of nitrogen is ammonia; a compound of nitrogen with hydrogen. If we act on this compound with chlorine, the chlorine takes the hydrogen (forming hydrochloric acid) and the nitrogen is set free, according to the equation,

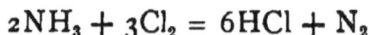

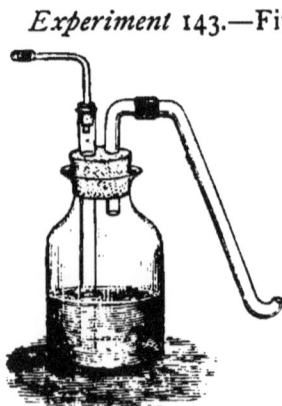

$$2NH_3 + 3Cl_2 = 6HCl + N_2$$

Experiment 143.—Fit a wide-mouthed bottle with a cork carry-

FIG. 83.

ing two wide glass tubes, as shown in Fig. 83. Half fill the bottle with strong ammonia solution, and pass a stream of chlorine (prepared as in Exp. 110) through the liquid. Notice that, as each bubble of chlorine enters the ammonia, there is a flash of fire in the liquid, so energetic is the chemical action that goes on. Observe also the white fumes; these consist of ammonium chloride; for, as soon as hydrochloric is formed (as shown in the equation), it combines with some of the ammonia in the bottle and forms ammonium chloride. Collect the gas which passes out, in the pneumatic trough. The object of the very wide tubes is that they may not get stopped up by the ammonium chloride. [Caution—do not let the experiment go on too long; as soon as one or two jars of gas have been collected, stop the operation.]

Another compound from which nitrogen can be readily obtained is ammonium nitrite, NH_4NO_2. When a strong solution of this salt is gently heated, it splits up into nitrogen and water.

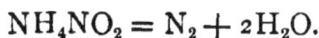

$$NH_4NO_2 = N_2 + 2H_2O.$$

Experiment 144.—Put about equal quantities of ammonium chloride and sodium nitrite into a flask, fitted with a cork and delivery tube, and about one-third fill the flask with water. Gently heat the mixture and collect the gas over water. As soon as the action begins, remove the lamp, and allow it to continue by itself. If the liquid begins to boil up too much, cool the flask by bringing a dish of cold water under it.

When a mixture of ammonium chloride and sodium nitrite is heated, the two salts interact on each other and give sodium chloride and ammonium nitrite, thus—

$$NH_4Cl + NaNO_2 = NaCl + NH_4NO_2,$$

and the ammonium nitrite then decomposes according to the first equation. The final result, therefore, of heating these two salts together is expressed thus—

$$NH_4Cl + NaNO_2 = NaCl + 2H_2O + N_2.$$

Properties of Nitrogen.—Nitrogen is very different from any of the gases we have as yet studied. When a taper is put into it we see that the gas is not like hydrogen, for it will not burn. It is not like oxygen, for it will not allow the taper to burn, but at once extinguishes it. It is not like chlorine or hydrochloric acid. The only property of this gas that can easily be shown is, that it seems to have no properties. It does not burn. It does not support combustion. It is not acid; does not bleach; does not act on metals; is not poisonous. Indeed nitrogen is one of the most inactive or inert substances we know. It will not support animal life, not because it is in any way injurious, but simply because

N

animals *must* have free oxygen to breathe; an animal placed in nitrogen dies from suffocation, just as it would if immersed in water.

Nitrogen is slowly absorbed by red hot magnesium, forming a compound of nitrogen and magnesium. This is one method by which nitrogen is separated from argon, which is even more inert a substance than nitrogen itself.

Although nitrogen will not burn, it will combine with oxygen slowly, when electric sparks are passed through a mixture of the two gases. In this respect again it differs from the still more inactive gas argon, which does not unite with oxygen when sparked with that gas.

EPITOME.

Nitrogen occurs uncombined in the air, to the extent of about four-fifths. It is obtained from air by withdrawing the oxygen, either by burning phosphorus or red hot copper. These combine with the oxygen and leave the nitrogen.

The chemical compound from which nitrogen is prepared, is ammonium nitrite. This when heated gives only water and nitrogen.

Nitrogen is characterised by great inertness. It is a colourless, odourless, tasteless gas ; does not burn, nor support combustion or respiration ; is not poisonous.

It does not easily unite directly with other elements. At a high temperature it combines with a few metals, and also with oxygen.

Just as chlorine is a member of a certain little family of elements (see *The Halogens*, p. 133), so nitrogen is also the representative of another group or family, which consists of the five elements *nitrogen, phosphorus, arsenic, antimony*, and *bismuth*. These have very little likeness to each other in their outward appearance, but they are closely related in their chemical behaviour.

Nitrogen and phosphorus are true non-metals, one being a gas, the other a wax-like solid ; they have, therefore, no properties which belong to metals, no metallic lustre, no power of conducting heat or electricity ; and they yield oxides which are *acid-forming*

Antimony and bismuth are metals ; they have metallic lustre, conduct heat and electricity, and form oxides which are *basic*.

Arsenic stands on the border line between the metals and non-metals, and is called a *metalloid*. It is a black shiny solid, with about as much lustre as graphite. It conducts heat and electricity, but its oxides are *acid-forming* compounds.

The detailed study of phosphorus, arsenic, antimony, and bismuth does not come within the scope of this elementary book.

Compounds of Nitrogen.—Although nitrogen *in the free state* is such an inert element, it has very strong chemical affinities when in combination with other elements. It forms a number of compounds with oxygen, with hydrogen, and with both oxygen and hydrogen together. It is also one of the constituents of a large number of animal and vegetable substances, where it is associated with carbon, hydrogen, and oxygen.

When such animal substances decay or "go bad," one of the first products of the decomposition is ammonia. The nitrogen in the compound combines with some of the hydrogen and forms this compound. Hence there is generally a strong smell of ammonia in stables and in urinals, where nitrogenous animal matter is undergoing decomposition. This natural process of decomposition is imitated artificially, when we heat such a compound so as not to allow air to get to it, so that it does not take fire. For example, when coal is heated in retorts, as in the manufacture of ordinary coal-gas, air does not get to the coal, and therefore it does not burn as it would in an open fireplace, but it is decomposed into a great variety of compounds, some solid, some liquid, and some gaseous. The process of heating substances in this manner is called *destructive distillation*, to distinguish it from the ordinary operation of distilling where the original substance is not decomposed. Now, although coal is very largely composed of carbon, it also contains amongst other constituents some nitrogen, and some hydrogen, and therefore, when it is destructively distilled, one of the gaseous products formed is ammonia. This dissolves in the water, which is another product of the decomposition, yielding the so-called "ammoniacal liquor" of the gas-works. This is the chief source from which ammonia is now obtained.

Another substance which is formed when animal matter containing nitrogen is allowed to decompose slowly by itself

in the presence of air, is nitric acid. Sometimes in the neighbourhood of ill-drained stables or dwellings, especially in hot
climates, where decomposing animal matter soaks into the
earth, crystals are to be seen on the soil or lower parts of the
walls.   These crystals consist of *nitre.*   The nitric acid formed
by the decomposition of the organic matter, combines with
potash present in the soil and forms potassium nitrate (*saltpetre*
or *nitre*).

At one time all our supplies of nitre were obtained by this
process, which was carried on by purposely mixing manure
and such decomposing refuse with wood ashes (which contain
a large quantity of potash) and earth, and allowing the heaps
to remain exposed to the air, occasionally moistening them
with drainage from manure.

Nowadays this process is not much used, because enormous
beds of sodium nitrate (called *Chili saltpetre*) have been found,
from which potassium nitrate can easily be made.

Ammonia and nitre are two of the most important compounds of nitrogen.

Ammonia.—This substance is a gas, but long before it
was known to be a gas, a solution of it in water was known,
and was called *spirits of hartshorn.*   It got this name because
it was obtained by the destructive distillation of horns and
hoofs of animals.   At the present day the source of all our
ammonia is the ammoniacal liquor of the gas-works.   This
is practically a solution of the gas in tarry water.   The liquid
used in the laboratory, and called ammonia, is simply a
solution of the gas in pure water.   If we heat such a solution
the gas is all expelled, and can be collected.

Experiment 145.—Gently heat a little strong solution of ammonia
in a flask provided with a cork and short exit tube bent at right
angles, and attach this to the apparatus for collecting gas by
upward displacement (Fig. 84).   Notice that the ammonia solution
gives off its gas so rapidly that it appears to be boiling, although
scarcely warmed to the temperature of the hand.   In a few minutes
the cylinder will be filled with ammonia gas.   Now remove it and
place it mouth downwards in a trough of water.   Notice that the
water quickly rises in the cylinder, showing that the gas is far

too soluble in water to allow of its being collected over that liquid.

Experiment 146.—Collect another cylinder of the gas, and as the glass is filling, hold a piece of litmus paper which has been reddened (by being dipped into very dilute acid) against the exit tube; observe that the gas is strongly alkaline. Also let the gas blow against a moistened piece of turmeric paper, and obtain the reddish brown stain due to the action of an alkali upon it. In the old days when all gases were called *airs*, ammonia was distinguished as the *alkaline air*. Cautiously smell the gas, *not by applying the nose to the tube*—this would give too strong a sniff of it and would be dangerous—but by gently wafting the escaping gas towards the face with the hand.

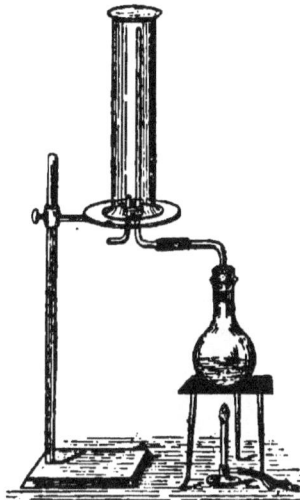

FIG. 84.

Gently thrust a lighted taper up into the gas. Carefully note the behaviour of this gas towards combustibles. When the taper is plunged right into the gas, the flame is extinguished, and the ammonia does not take fire. But when the taper flame is cautiously brought into the gas, at first the gas seems to be *trying* to burn; a curious brownish-yellow flame appears to surround the taper flame for a moment before the latter is put out.

Ammonia, therefore, will not burn in ordinary air, although it seems very nearly to do so. But if we add a little oxygen to the air, then the ammonia will burn quite easily.

Experiment 147.—Close one end of a wide glass tube (a gas-lamp chimney) with a cork through which two tubes pass: a moderately wide one reaching to the top of the chimney, and a narrow one passing just through the cork, as in Fig. 85. The wider of these tubes is attached to a small flask in which strong ammonia solution is gently heated; the other is connected to a supply of oxygen. The apparatus is supported in a clamp on a retort stand, not shown in the figure. A little plug of cotton wool should be pushed down to the bottom of the chimney, so as to cover the open end of the narrow tube through which the oxygen enters: this distributes the oxygen all round the centre pipe.

First regulate the little gas flame, so that a gentle stream of ammonia escapes up the centre tube. Then bring a lighted taper to the end of this tube and again notice the appearance of flame.

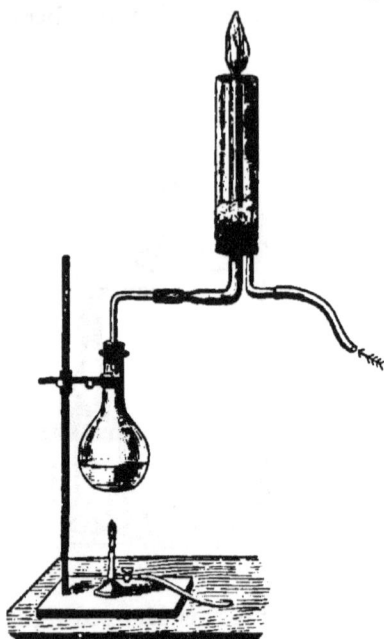

Now gently admit oxygen through the narrow tube, still holding the taper to the escaping jet of ammonia. As soon as a little oxygen reaches the top of the chimney the ammonia will light, and will continue burning without the taper, giving a curious yellow-brown coloured flame. Now stop the oxygen, and notice the ammonia flame gradually languish and go out.

Combination of Ammonia with Acids.

Experiment 148.—Pour a little dilute hydrochloric acid, dilute nitric acid, and dilute sulphuric acid into three separate little beakers, and add two or three drops of litmus solution to each. Now heat some strong ammonia solution in a flask, and pass a stream of the gas into each of the acids until the red colour of the litmus changes to blue, and then evaporate each solution to dryness in separate dishes heated gently over small rose burners. In each dish a residue is left, which but for the slight colour due to the litmus would be white. These residues are salts of ammonia; they consist of ammonium chloride (sometimes known by its ancient name of *sal-ammoniac*), ammonium nitrate, and ammonium sulphate.

FIG. 85.

Ammonium chloride and ammonium sulphate are commercially obtained by driving off the ammonia from the "ammoniacal liquor" of the gas-works, and passing the gas into either hydrochloric or sulphuric acid, just as in Exp. 148.

The equations which represent the combination of ammonia with these three acids are as follows:—

(1) $NH_3 + HCl = NH_4Cl.$
(2) $NH_3 + HNO_3 = NH_4NO_3.$
(3) $2NH_3 + H_2SO_4 = (NH_4)_2SO_4.$

How to get Ammonia out of its Salts.

Experiment 149.—Heat a small quantity of the ammonium chloride obtained in Exp. 148, in a dry test-tube. Note that the salt sublimes. Apply a lighted taper to the mouth of the tube, and see that it shows no signs of a flame of ammonia. Smell the tube, and hold reddened litmus paper to it, and observe not even a trace of ammonia is given off. *Therefore we cannot obtain ammonia from this compound simply by heating it.*

Experiment 150.—Heat a little ammonium nitrate in the same way. Note at once a great difference in the behaviour of this salt. It melts ; the chloride did not. Presently it effervesces ; evidently gas is coming off. Is this gas ammonia? Test with litmus ; note no blue effect. Smell the gas ; there is no smell of ammonia. Bring a lighted taper into the gas ; notice that the gas behaves like oxygen. Test with a glowing splint of wood ; it relights in this gas. Can it be oxygen? We must examine this point later ; in the mean time, the experiment shows that we *do not get ammonia by simply heating this salt.*

Experiment 151.—Heat a small quantity of ammonium sulphate in a similar manner. Notice that, as with the nitrate, this salt also melts and gives off gas. Test with litmus and turmeric, and observe alkalinity. Smell the gas, and note that when ammonium sulphate is heated alone, ammonia is given off.

Experiment 152.—Now take a small quantity of each of the three ammonium salts in three test-tubes, and add to each about the same quantity of powdered lime and apply heat gently. Notice that in each case ammonia is given off, as indicated by litmus or turmeric paper, as well as by the smell.

These four experiments show that it is only certain ammonium salts which give off ammonia when heated alone, but that all of them yield ammonia when heated with lime. We can, therefore, use this latter method as a test as to whether a particular salt is an ammonium compound or not. In practice, when it is required to make such a test, we usually employ a solution of sodium hydroxide (caustic soda) instead of the lime, as it is rather more convenient to use.

Experiment 153.—Add a little caustic soda solution to a small quantity of say ammonium chloride in a test-tube, and gently warm the mixture. Smell the escaping gas, and hold moistened litmus or turmeric paper to the mouth of the tube. Note abundance of ammonia.

Ammonia Solution is made by dissolving ammonia gas in water.

Experiment 154.—Place some ammonium chloride in a flask, and add about twice as much dry slaked lime and mix the two together. Arrange the flask in connection with a bottle containing water as shown in Fig. 86, and gently heat the flask. Air is first expelled, which bubbles through the water, but presently, as only pure ammonia escapes from the flask, the whole of the gas is absorbed by the water, no bubbles passing through the water. Notice that, after a while, the liquid in the bottle begins to get perceptibly warm to the hand. If, therefore, we wish to make a very strong solution of the gas, we must prevent this by immersing the bottle in cold water, for we have learnt by Exp. 145, that by warming a solution of ammonia the gas is expelled again. [Note. By arranging the apparatus so that the tube leading into the water is a good length, there is no fear of the liquid in the bottle being sucked back into the flask ; because if it begins to ascend far up the tube (which might happen if the evolution of gas were interrupted by a draught blowing the flame away for a moment), then air is at once drawn into the flask through the small quantity of mercury placed in the bend of the funnel tube.]

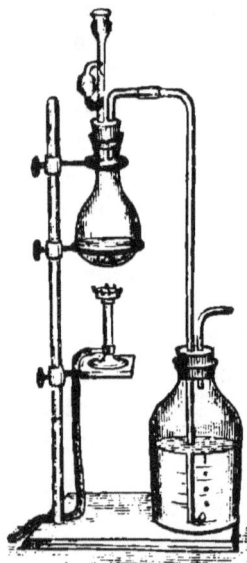

FIG. 86.

The chemical changes taking place when ammonium chloride is heated with caustic soda and with slaked lime, are expressed by the following equations :—

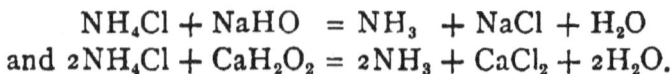

$$NH_4Cl + NaHO = NH_3 + NaCl + H_2O$$
$$\text{and } 2NH_4Cl + CaH_2O_2 = 2NH_3 + CaCl_2 + 2H_2O.$$

The nature of the change is the same in both cases, which

will be more easily understood if the equations be dissected in the following way. (1) With the sodium hydroxide—

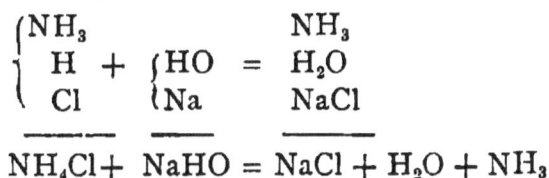

$$\left\{ \begin{array}{l} NH_3 \\ H \\ Cl \end{array} \right. + \left\{ \begin{array}{l} HO \\ Na \end{array} \right. = \begin{array}{l} NH_3 \\ H_2O \\ NaCl \end{array}$$

$$NH_4Cl + NaHO = NaCl + H_2O + NH_3$$

and (2) with slaked lime (or calcium hydroxide)—

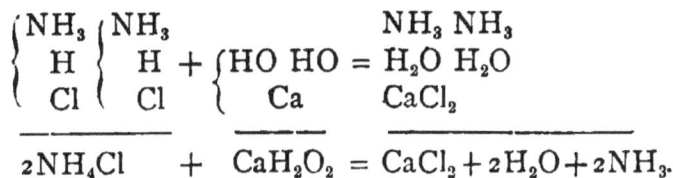

$$\left\{ \begin{array}{l} NH_3 \\ H \\ Cl \end{array} \right. \left\{ \begin{array}{l} NH_3 \\ H \\ Cl \end{array} \right. + \left\{ \begin{array}{l} HO \ HO \\ Ca \end{array} \right. = \begin{array}{l} NH_3 \ NH_3 \\ H_2O \ H_2O \\ CaCl_2 \end{array}$$

$$2NH_4Cl + CaH_2O_2 = CaCl_2 + 2H_2O + 2NH_3.$$

The Composition of Ammonia.—We have learnt that hydrogen and oxygen unite together very readily, forming water; also that hydrogen and chlorine combine with great ease, giving hydrochloric acid, so that we were able to gain knowledge of the composition of both water and hydrochloric acid by synthesis. But hydrogen only combines directly with nitrogen with great difficulty, and even then under exceptional conditions, therefore we cannot find the composition of ammonia by the direct union of its elements. We can, however, decompose ammonia, and find the proportional volumes of its constituents. We do this by means of chlorine, which has been shown (p. 176) to be able to decompose this compound, combining with the hydrogen and letting the nitrogen go free.

Experiment 155.—Take a long glass tube, closed up at one end, and divide it into three equal divisions with indiarubber bands. Fit to the tube a cork carrying a small stoppered dropping funnel, as shown in Fig. 87. Collect this long tube full of chlorine (using strong brine in the pneumatic trough) and insert the cork. Pour a little strong ammonia solution into the funnel, and allow it slowly to enter the tube, one drop at a time. The first few drops take fire as they go in (as in Exp. 143).

When about a dozen drops have been let in, the action is finished. Fill up the funnel with dilute sulphuric acid, so as to neutralize the

excess of ammonia, and fit a bent tube into the mouth of it, and let
the long end dip into a beaker of water as arranged in the figure.

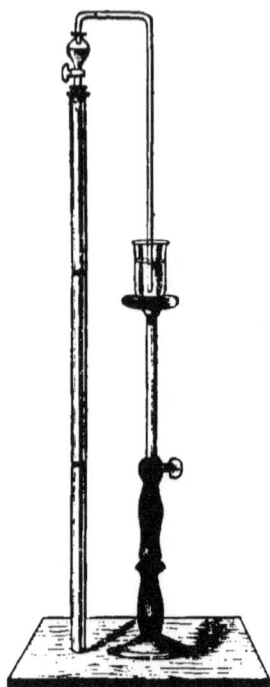

FIG. 87.

Then open the tap. Water at once runs
in, showing that some gas has disappeared,
and it will fill up exactly two of the measures,
leaving one measure of gas.

Test this gas by removing the cork and
dipping a lighted taper into the tube. The
gas is nitrogen. We therefore have got
one measure of nitrogen, which has been
expelled from combination with just so
much hydrogen as there was chlorine originally present in the tube, namely, three
measures (because chlorine and hydrogen
combine in equal volumes. See Exp. 103).
Therefore, in ammonia, the nitrogen and
hydrogen are combined in the proportion
one volume of nitrogen to three volumes of
hydrogen.

In order to prove that the formula
is NH_3 and not N_2H_6 (both of which
contain the two elements in the same
relative proportion), it is necessary to
find out the density of the gas by weighing a known volume
of it. When this is done, it is found that ammonia gas is
8·5 times as heavy as hydrogen. Its molecular weight, therefore, is twice this, namely, 17. This proves that the composition is expressed by the formula NH_3.

$$N = 14. \quad 3H = 3. \quad \therefore 14 + 3 = 17.$$

EPITOME.

Ammonia is formed by the putrefaction of organic matter containing nitrogen, such as manure. Minute quantities of it are
found in the air.

Ammonia is produced during the destructive distillation of coal
(as in the manufacture of ordinary coal gas), when it collects in the
watery liquid known as the "ammoniacal liquor."

Ammonia is obtained from its salts by heating them with slaked
lime or with sodium hydroxide (caustic soda)

Ammonia is a colourless gas, with a powerful pungent smell,

Ammonia.

and a strong alkaline reaction. It is very soluble in water, and therefore cannot be collected in the ordinary way. One litre of water at the common temperature dissolves about 800 litres of ammonia gas, while at 0° it will absorb as much as 1148 litres. All the gas is driven out of solution when the liquid is heated.

Ammonia will not burn in air, but it will burn in oxygen. It will not support the combustion of a taper. Ammonia is a light gas; it is just over one half as heavy as air, and is 8·5 times as heavy as hydrogen.

Ammonia is easily condensed to the liquid state. At the ordinary temperatures, about 7 atmospheres pressure will squeeze this gas into the liquid form. And again, if the gas, without being squeezed, is simply cooled below - 34° C. then it also turns into liquid ammonia.

[Note.—*Liquid ammonia* is not the same as a solution of ammonia in water.]

Liquid ammonia boils at − 33·7°, and has been largely used for the artificial production of ice.

When passed through a red-hot tube, or submitted to electric sparks, ammonia gas is decomposed into nitrogen and hydrogen. When a measured volume of the gas is thus decomposed, the volume is *doubled;* that is to say, two volumes of ammonia give one volume of nitrogen and three volumes of hydrogen.

The volume composition of ammonia is also established by decomposing it with chlorine, when it is found that three volumes of chlorine take three volumes of hydrogen to form hydrochloric acid, and leave one volume of nitrogen.

CHAPTER XX.

NITROGEN AND ITS COMPOUNDS (*continued*).

Nitric Acid, HNO$_3$.—This substance is one of the most important of all the nitrogen compounds, and from it, either directly or indirectly, we obtain all the other compounds of nitrogen which it will be necessary for us to study. Nitric acid is composed of nitrogen, hydrogen, and oxygen, but we do not make it by causing these three elements to unite directly. It *can* be produced synthetically, however, for if we add a little nitrogen to a mixture of oxygen and hydrogen, and ignite the mixture, the water which results from the union of the hydrogen and oxygen will be found to be *acid*, from the presence of a little nitric acid. Cavendish noticed this when making his experiments on the composition of water (p. 73), and was for a long time puzzled to account for the acidity of the water he got, but discovered that nitrogen from the air had found its way into the apparatus.

If a rapid stream of electric sparks is passed between platinum wires in a confined space of air, the oxygen and nitrogen begin to combine, forming an oxide of nitrogen. This gas has a brownish colour, and if a little water be then added the brown gas disappears; it dissolves in the water and yields nitric acid. In this manner nitric acid is produced during thunderstorms. The lightning flashes (which are simply enormous electric sparks) passing through the air cause the combination of some of the nitrogen and oxygen, and the oxide so formed is washed out by the rain.

Preparation.—Nitric acid is made from either potassium

nitrate (*nitre* or *saltpetre*) or from sodium nitrate (*Chili saltpetre*).

Experiment 156.—Place 50 grams of nitre in a glass retort with a stopper, and pour upon it the same weight of strong sulphuric acid. Gently heat the mixture and collect the distillate in a clean flask, which is kept cool by being placed in a dish of cold water, as shown in Fig. 88. Now and then turn the flask round, so as to cool it all over. Notice that reddish fumes appear in the retort, and

FIG. 88.

that a liquid collects in the flask which has a pale yellowish colour.

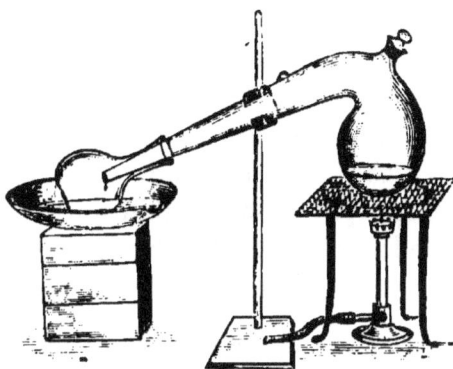

The equation which takes place in this experiment is the following—

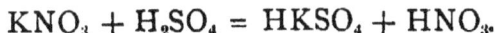

$$KNO_3 + H_2SO_4 = HKSO_4 + HNO_3.$$

The potassium in the nitre changes places with one-half of the hydrogen in the sulphuric acid, giving a salt called *hydrogen potassium sulphate*, and nitric acid. The salt remains behind in the retort.

If sodium nitrate had been used the reaction would have been quite similar.

The manufacturer of nitric acid always employs the Chili saltpetre, because it is a cheaper article than the potassium salt. He also carries on the operation at a higher temperature, using large cast-iron vessels in the place of a glass retort, which enables him to get all the hydrogen in his sulphuric acid exchanged for sodium, according to the equation—

$$2NaNO_3 + H_2SO_4 = Na_2SO_4 + 2HNO_3.$$

The same quantity of sulphuric acid, therefore, is made to decompose *twice* as much of the nitrate as in the laboratory experiment. At the high temperature, however, required to completely bring about this second reaction, a little of the nitric acid itself is decomposed, and therefore wasted.

Properties of Nitric Acid.—When quite pure, the acid is colourless. The sample obtained in Exp. 156 is coloured slightly yellow, because it has dissolved some of the coloured fumes which were produced.

Nitric acid fumes strongly when exposed to moist air. It can be mixed with water in any proportions. The strong acid is highly corrosive, and *must be handled with great care;* a few drops spilt upon the skin will cause bad wounds. Even when moderately diluted with water it will burn the clothes, and stain the skin yellow. If strong nitric acid is boiled, it begins to

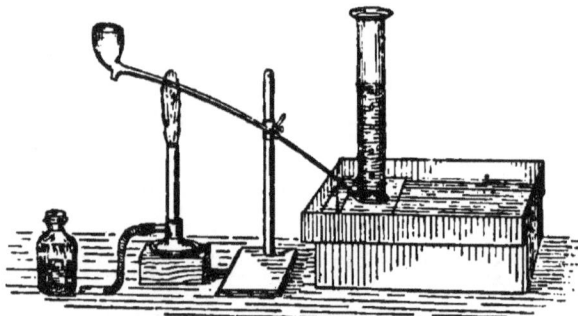

FIG. 89.

decompose, giving **oxygen** and **nitrogen peroxide** (a reddish-brown gas). When heated strongly this decomposition is very rapid.

Experiment 157.—Arrange a tobacco-pipe as shown in the figure, the mouthpiece just dipping beneath the water in the trough. When the stem is red-hot, pour a few drops of strong nitric acid into the bowl. As the acid passes the heated place it is decomposed into the two gases, oxygen and nitrogen peroxide. The latter gas, however, is very soluble in water, and therefore dissolves in the trough, while oxygen alone is collected.

Notice that, as the bubbles first appear in the water, they have a dark reddish colour, while the gas which actually collects is colourless. Test the gas for oxygen. The equation is the following—

$$2HNO_3 = H_2O + 2NO_2 + O.$$

Nitric acid is therefore a powerful oxidizing material, and it acts on many substances with great energy.

Experiment 158.—Gently heat a small quantity of sawdust in a small porcelain dish or crucible, until the wood has just become charred. Then cautiously let two or three drops of strong nitric acid fall upon the charred mass, and notice that it instantly takes fire. The charcoal burns at the expense of the oxygen supplied by the nitric acid, while fumes of the brown gas are produced at the same time.

Experiment 159.—Add a little powdered sulphur to a small quantity of strong nitric acid in a test-tube, and gently heat the mixture. Notice that the brown-coloured oxide of nitrogen is again given off in quantity. The acid is therefore being *reduced;* in other words, it is giving up some of its oxygen to the sulphur, which is gradually being converted into sulphuric acid. Let the action continue for a few minutes, and then test the liquid for sulphuric acid in the following way—

Add one or two drops of sulphuric acid to some water in a test-tube, and then add some solution of barium chloride. Notice a white precipitate. The reaction here is

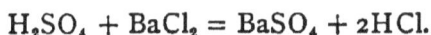

$$H_2SO_4 + BaCl_2 = BaSO_4 + 2HCl.$$

The barium and hydrogen change places, and barium sulphate is produced. This substance, being insoluble in water, separates out as a white solid. There are, however, other things which would give a white precipitate with barium chloride, but barium sulphate may be distinguished from the other precipitates by the fact that it cannot be dissolved by acids. Therefore, as a confirmation of the test, pour half the white precipitate into a second test-tube, and to one portion add some strong hydrochloric acid, and to the other some nitric acid, and note that the precipitate is not dissolved in either case.

Now apply this test in the case of the nitric acid in which sulphur has been heated. Dilute the acid with water, and add a few drops of barium chloride. A white precipitate proves the presence of sulphuric acid, which requires no confirmatory test, as the liquid already contains nitric acid.

Besides oxidizing such elements as carbon, sulphur, phosphorus, iodine, etc., it attacks a great many metals, converting them into compounds which contain oxygen, and are, therefore, oxidation products.

Experiment 160.—Drop a few fragments of copper wire or foil into a little nitric acid in a test-tube. Notice that violent action at once sets in. Torrents of the red gas are given off, and the copper quickly disappears, while the liquid becomes blue. The copper is converted into copper nitrate (a blue salt), which remains dissolved in the liquid. A number of other· metals behave in a similar way, such as silver, mercury, iron, lead, zinc. They are converted into nitrates, while the nitric acid is reduced to one of the oxides of nitrogen.

We may suppose that the first action of the acid on such metals is that the metal changes place with the hydrogen, thus—

$$Cu + 2HNO_3 = H_2 + Cu(NO_3)_2,$$

and that immediately this nascent [1] hydrogen attacks another portion of nitric acid, taking away some of its oxygen, whereby water is formed and an oxide of nitrogen left. *Which* oxide of nitrogen is left depends on a number of circumstances, for hydrogen in this state can gradually take *all* the oxygen from nitric acid, leaving at last only nitrogen. *In no case by the action of nitric acid on metals is any hydrogen given off so as to be collected.* On account of its powerful action in dissolving metals, nitric acid used to be called *aqua-fortis* (the strong water) ; but even this acid cannot dissolve the " noble " metals, gold and platinum.

Aqua Regia.—Both nitric acid and hydrochloric acid are without any action on gold or platinum, but if the two acids are mixed together, then the mixture will easily dissolve either of these metals.

Experiment 161.—Place a gold leaf in a wide test-tube and pour strong nitric acid upon it. The leaf breaks up into small particles, but notice that it is not dissolved, and may be left a long time in the acid. Do the same with hydrochloric acid in another test-tube, and notice that, in like manner, the gold leaf is not acted upon. Now mix the contents of the two tubes, and in a few moments the gold will be entirely dissolved.

The mixture of these two acids is known as *aqua regia* (the royal water), just because it is able to dissolve the noble metals. When metals dissolve in aqua regia, it is always the chloride, and not the nitrate of the metal, that is formed.

[1] Nascent means *just born ;* and elements at the moment of their liberation from combination, are said to be in the *nascent state ;* at this particular moment the element is much more chemically active.

Impurities in Nitric Acid.—Common nitric acid generally contains a number of impurities, the most usual of which are sulphuric acid (derived from the sulphuric acid used in its manufacture) and iron (from the vessels in which the acid is made). The sulphuric acid may be tested for by the method described in Exp. 159. The presence of iron in the acid may be found out by the following test.

Experiment 162.—Dilute a little of the suspected acid, and add to it a few drops of a solution of potassium ferrocyanide (*yellow prussiate of potash*). If much iron is present, a deep blue precipitate will be seen (" Prussian blue ") ; but if there is only a little of this impurity in the acid, the ferrocyanide will only produce a bluish or greenish colour.

Nitrates.—The salts which nitric acid forms, when its hydrogen is replaced by metals, are called *nitrates ;* such as potassium nitrate, copper nitrate, etc.

Like hydrochloric acid, nitric acid has only one atom of hydrogen in it ; it is on this account called a *mono-basic acid.* Nitrates are not only produced by dissolving metals in the acid, but also by acting on metallic oxides, hydroxides, or carbonates, with nitric acid.

Experiment 163.—Place in three separate dishes a little copper oxide, potassium hydroxide, and sodium carbonate. Add a little strong nitric acid to the copper oxide and gently warm it. Notice that the oxide dissolves, giving the same blue solution of copper nitrate as in Exp. 160, but that no brown gas is given off.

Dilute some nitric acid, and add it gradually to the potassium hydroxide and the sodium carbonate until each is dissolved. Now slowly evaporate all three solutions down to dryness. Blue crystals of copper nitrate, and white crystals of potassium and sodium nitrates will be obtained. The following are the three equations for the formation of these nitrates—

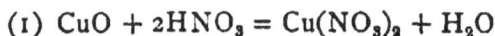

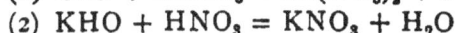

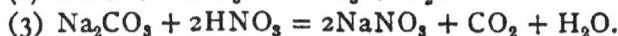

(1) $CuO + 2HNO_3 = Cu(NO_3)_2 + H_2O$
(2) $KHO + HNO_3 = KNO_3 + H_2O$
(3) $Na_2CO_3 + 2HNO_3 = 2NaNO_3 + CO_2 + H_2O.$

Experiment 164.—Spread a drop or two of the solutions of sodium nitrate and potassium nitrate obtained in the last experiment, upon separate pieces of clean, flat glass, and allow the solution to evaporate by itself. Carefully compare the crystals of the one salt with those of the other, using, if necessary, a pocket lens. The shapes of the crystals are quite different. Those of potassium nitrate are long thin prisms, while the sodium nitrate crystals are more like little cubes ; although not exactly cubes. [Sodium nitrate is sometimes called *cubical nitre* for this reason.]

All nitrates, when strongly heated, are decomposed, and, in most cases, they give off oxygen. They are, therefore, like nitric acid itself, powerful oxidizing agents, and will readily give up oxygen to substances capable of taking it.

Experiment 165.—Heat a few crystals of potassium nitrate in a test-tube. They first melt, and presently give off gas. Now drop into the test-tube a fragment of charcoal, about the size of a grain of corn. The charcoal takes fire when it touches the hot nitre, and the little fragment dances about on the molten salt. Next drop in a piece of sulphur about the same size, and note how readily it burns in contact with the melted nitre.

These three substances, nitre, charcoal, and sulphur, are what gunpowder is composed of. The nitre affords a supply of oxygen sufficient to burn up the carbon and sulphur, so that these materials can burn in places where they cannot get oxygen from the air, such as in the breech of a gun, or even underneath water.

Experiment 166.—Weigh out 15 grams of finely powdered nitre, 3 grams of charcoal (also finely powdered), and 2 grams of flowers of sulphur. Mix these together most thoroughly, *not with a pestle and mortar*, but by putting them all into a sieve, and shaking them through together (a sieve may easily be made by tying a piece of muslin over the mouth of a beaker, the bottom of which has been broken).

Make a little heap with a portion of this mixture, upon a piece of wood, and set fire to it with a match or taper. Notice how it burns—not so suddenly as proper gunpowder burns, because the manufacturer is able to get the ingredients much more intimately mixed than in this case.

Experiment 167.—Make a little paper tube about 7 or 8 centimetres long (3 inches) by rolling a piece of writing paper closely round a lead-pencil about half a dozen turns. Fasten the edge down with a little strong gum. Slip the pencil out, and stop up one end of the paper tube with a tiny cork and a little sealing-wax. Now proceed to fill the tube with the mixture prepared in the last experiment, putting in a little at a time, and packing it tightly down, using the lead-pencil as a ram-rod. It is now much the same thing as an ordinary " squib," but without the " bang."

Now light the open end, and while it is burning plunge it under water, notice that it continues burning, and gives off a quantity of gas which bubbles up through the water.

Tests for Nitrates.—All nitrates are soluble in water, and the presence of such a salt in a solution can easily be detected by either of the two following tests.

Experiment 168.—Dissolve a crystal of potassium nitrate in a little water, and to a part of the solution so obtained add some strong sulphuric acid. The nitrate is converted into free nitric acid (just as in the preparation of nitric acid. Exp. 156). Then drop into the mixture a fragment or two of metallic copper. As we have learnt, nitric acid acts on copper, and red fumes are evolved ; so that if the liquid containing the copper be now gently warmed, red fumes will be given off, which proves that a nitrate was originally present.

The second test is more delicate, and will detect much smaller quantities of a nitrate. Take another portion of the solution containing the nitrate, and, as before, add sulphuric acid. Now pour gently down into the test-tube a solution of ferrous sulphate, so that this solution floats on the top of the other. Notice that where the two liquids meet in the tube, there is a dark brown layer formed ; which, if the ferrous sulphate has been added quite carefully so as not to get mixed with the other solution, will appear as a well defined ring.

[The nitric acid produced by the addition of sulphuric acid to the nitrate, gives up some of its oxygen to the ferrous sulphate, and a small quantity of one of the oxides of nitrogen is formed. This dissolves in a further portion of ferrous sulphate, forming a dark brown compound.]

EPITOME.

Nitric acid (*aqua fortis*) is made by the action of sulphuric acid on potassium or sodium nitrate. On the manufacturing scale sodium nitrate is used (*Chili saltpetre*), as this salt is cheaper. Nitric acid is produced when electric sparks pass through air in the presence of water, hence it is formed during thunderstorms, and gets washed into the ground with the rain.

Pure nitric acid is a colourless fuming corrosive liquid.   It begins to decompose when boiled, giving red fumes of nitrogen peroxide, and oxygen.

Nitric acid attacks most metals, but has no action on gold or platinum.   The result of the action of nitric acid on metals is either a nitrate or an oxide of the metal, and one or more of the oxides of nitrogen : but *hydrogen is never evolved.*

We can at once distinguish gold from any imitations of this metal, by touching the surface with a drop of nitric acid.   If there is no action, the material is gold.

Nitric acid is a powerful oxidizing substance, and converts the elements sulphur, phosphorus, and iodine into sulphuric, phosphoric, and iodic acids respectively, with evolution of oxides of nitrogen.

The salts of nitric acid are *nitrates.*   Potassium nitrate (*nitre*, *saltpetre*) is one of the most important.   It is a constituent of gun· powder.   When heated, nitrates decompose and give up oxygen ; sometimes forming first of all a *nitrite*, thus—

$$KNO_3 = O + KNO_2$$

Nitrates of certain metals, when heated, leave an oxide of the metal, and give off nitrogen peroxide and oxygen, thus—

$$Pb(NO_3)_2 = PbO + 2NO_2 + O.$$

CHAPTER XXI.

THERE are five oxides of nitrogen; their names and formulæ are as follows :—

(1) Nitrous oxide (*laughing gas*) ... N_2O
(2) Nitric oxide ... ... NO
(3) Nitrogen trioxide ... ... N_2O_3
(4) Nitrogen peroxide ... ... NO_2
(5) Nitrogen pentoxide ... ... N_2O_5

Numbers 1, 2, and 5 are acidic oxides. The acid derived from nitrous oxide (called *hyponitrous acid*) and that from nitrogen trioxide (*nitrous acid*) are both very unstable compounds, and have never been obtained in the free state. The acid derived from nitrogen pentoxide is *nitric acid.*

Nitric Oxide, NO.—It will be most convenient to study this oxide first.

When nitric acid acts on metals, as explained on page 192, the hydrogen which is first displaced from the acid by the metal, at once attacks a further portion of nitric acid, depriving it of more or less of its oxygen, and thereby reducing it to one or other of the oxides of nitrogen. The gradual reduction of nitric acid by hydrogen will be seen by the following equations :—

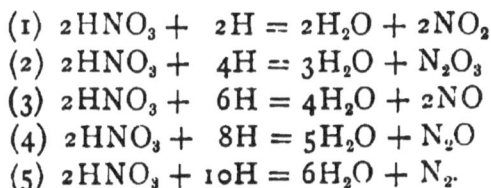

(1) $2HNO_3 + 2H = 2H_2O + 2NO_2$
(2) $2HNO_3 + 4H = 3H_2O + N_2O_3$
(3) $2HNO_3 + 6H = 4H_2O + 2NO$
(4) $2HNO_3 + 8H = 5H_2O + N_2O$
(5) $2HNO_3 + 10H = 6H_2O + N_2.$

The reducing action may even go a step further, and result in the formation of ammonia, thus—

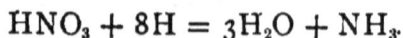

$$HNO_3 + 8H = 3H_2O + NH_3.$$

Now it depends partly on what metal is being acted on by the nitric acid, partly on the strength of the acid used, partly on the temperature, and partly on the amount of the nitrate of the metal which is produced during the action, which particular oxide will be produced.

For instance, when nitric acid is poured upon copper, at first some nitrogen peroxide is formed, then nitric oxide is evolved. After a time nitrous oxide begins to come off, and later on nitrogen is formed. Probably at no single moment is any one oxide *alone* produced, so that we do not get any one quite free from the others by such an experiment. In many cases, however, we know the particular conditions which will give us the oxide we want in a state which is sufficiently free from the others for ordinary experiment.

Thus, in order to get nitric oxide, we use copper and nitric acid, doing the experiment in the following way.

Experiment 169.—Place a quantity of copper clippings in a two-necked bottle, arranged as in Fig. 39, p. 45. Pour upon the copper a considerable quantity of a mixture consisting of one part of strong nitric acid and two parts of water (by measure). In a few minutes a brisk action sets in, and for some little time the gas which is evolved consists *chiefly* of nitric oxide. Collect three or four jars of the gas as quickly as possible, and then empty out the apparatus. *Avoid breathing the gas, as it is injurious.*

The reaction which expresses the production of nitric oxide, is that given in equation No. 3 above, where six atoms of hydrogen act on two molecules of nitric acid, the six atoms of hydrogen being displaced from six molecules of nitric acid. The two steps or stages in the reaction may be expressed by the two equations—

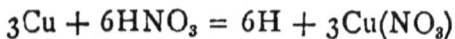

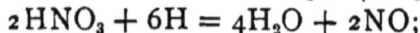

$$3Cu + 6HNO_3 = 6H + 3Cu(NO_3)$$
$$2HNO_3 + 6H = 4H_2O + 2NO;$$

Or to express the final results by a single equation—

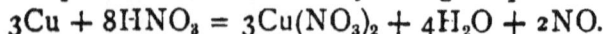

$$3Cu + 8HNO_3 = 3Cu(NO_3)_2 + 4H_2O + 2NO.$$

As already stated, other reactions resulting in the formation of other oxides of nitrogen are liable to go on at the same time as this one; hence when chemists want perfectly pure nitric oxide, they make it by other methods.

Properties of Nitric Oxide.—The samples collected show that it is a colourless gas, which is not appreciably soluble in water. Its most characteristic property is its behaviour when it comes in contact with the air.

Experiment 170.—Take one of the jars of gas out of the trough, and for a moment uncover its mouth. Notice that where the gas and air meet, dark red fumes are formed. [This jar may be returned to the trough, as it will do for another experiment.]

Which of the constituents of the air is the cause of these red fumes?

Experiment 171.—Decant some of the gas from one of the jars into a narrow cylinder filled with water and standing in the trough, until the cylinder is about three parts filled with gas. Now bubble up into this a little oxygen [either from another cylinder, or better from a bottle of compressed gas]. Notice that as each bubble of oxygen comes into the nitric oxide, it forms the same red gas as before. Therefore, it was the oxygen in the air which caused it in the former experiment.

Notice that, as each addition of oxygen is made, there is a momentary expansion, due to the heat produced by the combination of the two gases; and also that, after a moment, the volume diminishes and the red gas disappears.

The compound that is formed is nitrogen peroxide, NO_2. This is the red gas, and it is a gas which is quickly dissolved by water. If, therefore, the sample of nitric oxide is quite pure, and oxygen is added cautiously with frequent shaking up with the water, the entire quantity of gas ought to disappear altogether, being wholly converted into nitrogen peroxide, which will dissolve in the water.

Since nitric oxide combines with atmospheric oxygen the moment it mixes with the air, of course we cannot tell whether the gas has any smell, as before we could smell it, it is no longer nitric oxide, but nitrogen peroxide; and this gas has a most unpleasant smell and is poisonous.

Next try the action of combustibles on nitric oxide.

Experiment 172.—Introduce a lighted taper ; note that the gas
does not burn, and that it at once puts out the taper.

Place a small piece of phosphorus in a deflagrating spoon ; set
fire to it, and before it has time to burn up, plunge it into a jar of
the gas. The phosphorus will be extinguished. Now withdraw
it, and allow it to burn up brightly before putting it into the gas ;
note that now it burns in the gas with a bright flame, and forms
white fumes, like those produced when it burns in air or oxygen.

This shows that nitric oxide itself does not support the
combustion of common combustibles, but that if the burning
substance is hot enough to decompose the gas into nitrogen
and oxygen, then combustion goes on in the oxygen so
liberated. Therefore, the product of burning is the same as
in the air, but as there is a larger proportion of oxygen in
nitric oxide than in air, the combustion is more rapid and
brilliant.

The composition of nitric oxide is found by taking a
measured volume of the gas, and strongly heating in it some
metal, such as iron, which can combine with the oxygen, and
leave the nitrogen. When this is done, it is found that the
volume of the nitrogen left is exactly half the volume of the
original gas. This, however, does not tell us anything about
the volume of the oxygen in the compound unless we know
the *density of nitric oxide.* When the gas is weighed, it is
found to be 15 times as heavy as hydrogen. Two litres of it
therefore weigh 30 criths. But as nitric oxide contains half
its volume of nitrogen, in these two litres there must be one
litre of nitrogen. Now one litre of nitrogen weighs 14 criths,
therefore we get—

2 litres of nitric oxide, weighing 30 criths
contain 1 litre of nitrogen, weighing 14 „
—————
leaving 16 criths as the weight of
the oxygen ;

but 16 criths is the weight of 1 litre of oxygen, therefore 2
litres of nitric oxide contain 1 litre of nitrogen and 1 litre of

oxygen ; or, in other words, 2 volumes of nitric oxide contain 1 volume of nitrogen and 1 volume of oxygen.

Nitrogen Peroxide, NO_2.—This is the red-brown gas which is formed when nitric oxide comes in contact with air or oxygen

$$NO + O = NO_2$$

It is also given off when many nitrates are heated.

Experiment 173.—Heat a little lead nitrate in a test-tube ; notice that the tube is soon filled with the dark, red-brown gas. We have already seen that this gas is very soluble in water, therefore we cannot collect it over water. Attach a cork and delivery tube to the test-tube, and see if any gas is coming off besides the nitrogen peroxide, by collecting it over water. Note that a colourless gas is collected. Test this gas with a glowing splint ; the gas is oxygen. Therefore, oxygen and nitrogen peroxide are both produced. The equation is—

$$Pb(NO_3)_2 = PbO + 2NO_2 + O$$

This gas plays an important part in the manufacture of sulphuric acid, as will be explained later on.

Nitrous Oxide, N_2O.—This compound is easily prepared by simply heating the salt ammonium nitrate.

Experiment 174.—Place some crystals of ammonium nitrate in a small flask fitted with a cork and delivery tube, and gently heat over a small flame. The crystals quickly melt and give off gas. Collect the gas over water in the pneumatic trough, using water *as warm as can be comfortably borne by the hands*. Collect several jars full. The equation is the following—

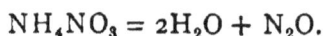

$$NH_4NO_3 = 2H_2O + N_2O.$$

[Compare this reaction with that by which nitrogen was produced by heating ammonium nitrite.]

Properties of Nitrous Oxide.—From the examples collected, it is seen to be a colourless gas. It has a faint, rather pleasant smell, and a sweetish taste. The gas is moderately soluble in cold water, but less so in warm ; therefore there is not so much loss of gas if warm water is used for its collection. When nitrous oxide is inhaled for a short time, it causes a kind of intoxication, often accompanied by boisterous laughter. On this account it is often called *laughing gas.* If the inhalation is continued it produces a state of insensibility to pain, and for this reason it is largely used by dentists.[1]

Experiment 175.—Bring a lighted taper into a jar of nitrous oxide. Notice that the gas behaves like oxygen, for it does not burn, but causes the taper to burn brightly. Dip a glowing splint of wood also into the gas ; the splint is rekindled ; so that from these experiments we could not distinguish this gas from oxygen. Burn a piece of phosphorus in the gas, using a deflagrating spoon as in Exp. 65. The phosphorus burns just as though the gas were oxygen.

How, then, can we distinguish between nitrous oxide and oxygen ?

Experiment 176.—Place a jar of oxygen and one of nitrous oxide side by side. Take a fragment of sulphur on a deflagrating spoon, or on a bundle of asbestos (see Exp. 64), light one corner of it in a gas flame, and before it has time to burn up at all, dip it into the oxygen. Notice that instantly it burns up vividly and continues burning. Now do the same in the other gas. Notice what happens ; the sulphur is extinguished. Repeat this once or twice in the same jar, to be quite sure of it. Now let the sulphur burn up well before plunging it into the gas, and notice that it continues burning when thrust into the nitrous oxide.

Nitrous oxide, then, differs from oxygen in that it will extinguish burning sulphur unless the sulphur is thoroughly hot when brought into it.

Another and very certain way of distinguishing between nitrous oxide and oxygen, is to mix each of them with nitric oxide. We have already seen what happens when nitric oxide is mixed with oxygen (Exp. 171). But when nitric oxide is mixed with nitrous oxide there is no production of brown fumes.

Nitrous oxide and oxygen can also be distinguished by mixing each of them with hydrogen, and exploding them. For instance, if a mixture of equal volumes of oxygen and hydrogen is exploded, we know from what we have learnt of these gases,

---

[1] Students are advised not to experiment upon themselves, or each other, by inhaling the gas.

that the hydrogen will combine with half the oxygen to form water, and that half the original volume of oxygen will remain over. But when a mixture consisting of equal volumes of nitrous oxide and hydrogen is exploded, the volume of gas left is the *same as the original volume of nitrous oxide*, and this gas is found to be *nitrogen.*

This last experiment also teaches us the composition of nitrous oxide. The volume of nitrogen present is equal to the volume of the nitrous oxide, and the oxygen present was just enough to unite with a quantity of hydrogen equal in volume to the nitrous oxide. We know that this amount of oxygen is exactly half the volume of hydrogen, therefore half the volume of the nitrous oxide. In other words, two volumes of nitrous oxide contain two volumes of nitrogen and one volume of oxygen.

Nitrous Acid and Nitrites.—Nitrous acid has never been obtained in a pure state, and even very dilute solutions of it quickly decompose.

The formula for the acid is HNO_2; it has one atom less oxygen than nitric acid, and it can readily take up oxygen from compounds which are rich in oxygen, and pass into nitric acid.

Experiment 177.—Dissolve a few particles of sodium nitrite in half a test-tube of water, and add to it a few drops of sulphuric acid. This decomposes the sodium nitrite, forming sodium sulphate and nitrous acid, which can exist for a short time in the dilute solution. Now add to this some solution of potassium permanganate (a salt very rich in oxygen), and notice how the purple colour of the permanganate is instantly destroyed, owing to the nitrous acid depriving it of some of its oxygen.

On the other hand, nitrous acid will give up some of its own oxygen to many substances which are ready to unite with oxygen.

The salts of this acid are called nitrites, as sodium nitrite, potassium nitrite, etc. Potassium nitrite can easily be made by gently melting potassium nitrate, when the nitrate parts with some oxygen and leaves the nitrite—

$$KNO_3 = KNO_2 + O.$$

We can easily tell a nitrite from a nitrate by adding dilute sulphuric acid to each. The nitrite at once evolves brown fumes, the nitrate does not.

EPITOME.

Five oxides of nitrogen are known.

1. *Nitrous oxide* (laughing gas), N_2O, is obtained by heating ammonium nitrate.

It is a colourless, slightly sweet tasting gas. Does not burn, but supports combustion almost as well as oxygen. Distinguished from oxygen by (*a*) much greater solubility in water, (*b*) gives no red fumes when mixed with nitric oxide.

Nitrous oxide is used by dentists to produce insensibility to pain. It is called laughing gas, because when a little of it mixed with air is inhaled it produces hysterical laughter.

2. *Nitric oxide*, NO, is prepared by acting on copper with nitric acid. Other oxides of nitrogen are produced at the same time, although at a certain stage of the action the bulk of the gas given off is nitric oxide. It is a colourless gas which instantly combines with free oxygen to form nitrogen peroxide. Therefore, when it comes into the air it at once forms brown fumes of the peroxide. Nitric oxide extinguishes a taper, because the flame is not hot enough to decompose the gas. Even phosphorus, unless strongly burning, is extinguished; but if burning vigorously when introduced into the gas, the flame is hot enough to decompose nitric oxide into oxygen and nitrogen, when, of course, the phosphorus will burn.

Nitric oxide is distinguished from all other gases by forming red fumes in contact with air or oxygen.

3. *Nitrogen trioxide*, N_2O_3, does not exist as a gas. When a mixture of nitric oxide, NO, and nitrogen peroxide, NO_2, in equal volumes, is passed through a strongly cooled tube, a blue liquid condenses, which is believed to be nitrogen trioxide.

4. *Nitrogen peroxide*, NO_2. This gas is formed by heating lead nitrate. When the gas is passed through a cooled tube, it condenses to a yellowish liquid, which is condensed nitrogen peroxide. The liquid gives off the familiar reddish brown gas.

When the gas is heated, it rapidly darkens in colour. This change is due to the molecules breaking up into simpler ones. At low temperatures nitrogen peroxide has the composition N_2O_4, its density being 46; but as the temperature gradually rises, the gas breaks up into molecules having the composition NO_2, and at 140° its density is 23. Therefore nitrogen peroxide has two formulæ, N_2O_4 at low temperatures, and NO_2 at high temperatures. Nitrogen peroxide dissolves in water, therefore can only be collected by displacement.

5. *Nitrogen pentoxide*, N_2O_5, is obtained by withdrawing the elements of water from pure nitric acid by means of phosphorus pentoxide.

$$2HNO_3 - H_2O = N_2O_5.$$

It is a white crystalline substance which cannot be preserved.

It combines with water with great eagerness, forming nitric acid.

Nitrous acid, HNO_2, is not known in a pure state. Its salts are nitrites.

CHAPTER XXII.

OZONE.

WHEN we hear of " Smith *alias* Sampson " getting into trouble, we know at once that the two names stand for the same person. Very often when a man assumes an *alias* he alters his outward appearance as far as he can in order to make it more difficult to recognize him. Perhaps from being a fair and beardless person, he now appears as a dark man with a black beard ; having a scar on his face, and perhaps even a squint.

Now some of the elements are capable of doing something of the same kind. Under certain circumstances they are able to assume an *alias*, as it were. They adopt new outward appearances and fresh properties so different from what they usually have, that they are then altogether unlike their original selves, and are quite likely to be mistaken for some entirely different element. Oxygen is one of the elements which can do this. With the usual properties of this element we are now quite familiar ; but when it assumes its *alias*, we find first that it adopts a powerful and unpleasant smell. Also it develops a habit of attacking organic substances in a violent manner, so that it cannot even be passed through a piece of indiarubber pipe without instantly destroying it. Also, it attacks metals like silver and mercury, which it took no notice of before ; and if it comes into contact with potassium iodide it instantly seizes it, steals the potassium from it, and turns the iodine adrift.

We see, therefore, that when oxygen adopts these new habits and properties, it seems like an entirely different substance, although all the time it is nothing else but oxygen.

Phosphorus is another element which can do the same thing. As usually seen, phosphorus is a wax-like solid, very poisonous, easily cut with a knife, melts when warmed by the hand, and takes fire so easily that it has to be kept beneath water, and care has to be taken in handling it. When it assumes its *alias*, however, it becomes a dark red substance, which looks like chocolate; it is no longer poisonous, does not melt, and requires to be strongly heated to make it burn. All the time it is phosphorus, and nothing else but phosphorus, but yet so unlike the element in its ordinary state.

The word that is used for this curious property is *allotropy* (meaning "other form"). We say that the element when appearing in its *alias*, or its more unusual character, is an *allotropic modification*, or more shortly, an *allotrope* of that element. Thus, when phosphorus appears as the red substance, we call this the allotropic modification of phosphorus, and when oxygen is in the condition in which it shows such active properties, we speak of it as allotropic oxygen, or ozone (ozone means "a smell").

We can make oxygen assume this allotropic state in several ways. The best method, and the one which gives the largest amount of ozone, is to expose oxygen to a particular kind of electrification, known as the *silent discharge*.

Experiment 178.—Take a piece of narrow glass tube about 30 centimetres (12 inches) long and wind a spiral of thin copper wire round the entire length outside : about three turns of wire to each centimetre.

Attach a piece of indiarubber tube to one end, so as to conduct a stream of oxygen through. Pierce a hole with a pin through the

FIG. 90.

rubber just beyond the end of the glass tube, and push a thin copper wire through it, so that the wire reaches nearly to the other end of the glass tube, as shown in Fig. 90.

The inside and outside wires are then connected to a Ruhmkorf's coil. When the coil is set in action, there will be no visible spark jumping across between the wires, but a vast number of tiny sparks passing from the whole length of one wire to the other. If now a slow stream of oxygen be passed through the tube, it has to "run the gauntlet" of this host of little sparks, and this causes a *part* of the oxygen to change itself into ozone. *Why* it does this, nobody knows.

Test for Ozone.—We test for ozone by making use of the property it has of setting free the iodine contained in potassium iodide; and in order that we may see when the iodine is so liberated, we have starch present. The instant the iodine is set free it unites with the starch and produces the blue compound described on page 134.

Experiment 179.—Make a little thin starch paste (see p. 134), and add to it a little potassium iodide solution. Cut strips of white paper and dip them into this mixture, and hang them up to dry. These are called ozone test papers.

Moisten such a test paper and hold it at the end of the tube, where ozone is being formed in the last experiment. Notice that the paper is instantly turned blue. This means that ozone has decomposed some of the potassium iodide, and that the free iodine has then united with the starch. Notice the curious smell of the ozone which passes out of the tube.

Ozone is quickly transformed back into ordinary oxygen by being heated.

Experiment 180.—Attach to the end of the "ozone tube" a short straight tube, by means of a rather wider tube and two corks,

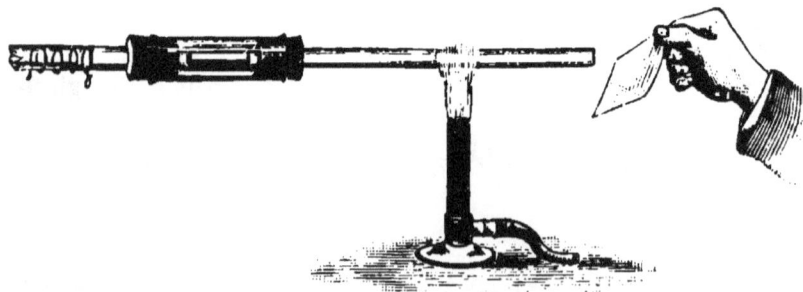

FIG. 91.

as shown in Fig. 91. (The usual indiarubber joint cannot be used, for the reason stated above.) Pass the ozonized oxygen through, and at the same time heat the tube. The ozone will be changed into ordinary oxygen, and if the gas passing out is tested with a piece of ozone test paper, there will be no blue colour produced.

Chemists now know that the change which oxygen undergoes in passing from oxygen into ozone, is that the molecule of oxygen takes up another atom of oxygen. Molecules of oxygen consist of two atoms, while molecules of ozone are composed of three atoms of the same element. Therefore, when oxygen changes into ozone there is a contraction in the volume, and *vice versâ*, when ozone passes back into ordinary oxygen there is an expansion in the volume. Ozone is on this account sometimes spoken of as " condensed oxygen." This is quite true in one sense, because the molecule of ozone is more condensed than the molecule of oxygen ; but it must be remembered that this is quite a different thing from condensing or compressing oxygen. We cannot condense oxygen into ozone merely by compressing the former so as to reduce the volume. Ozone is present in small quantities in country air. It is produced by lightning discharges.

It is also formed in small quantities when certain substances combine with oxygen without actually burning. Thus, if a piece of phosphorus is placed on the table, it is seen to "smoke." It is really oxidizing rather quickly, but is not actually burning. Now the phosphorus, in thus combining with atmospheric oxygen, causes a *little* of the oxygen to go into the allotropic form. How and why it does so, is not known.

Experiment 181.—Take a short stick of phosphorus (if it is coated over with a white or greyish film, scrape it clean, *underneath water*) and place it in a good large bottle, having a small layer of water on the bottom—just enough to half bury the phosphorus. Cover the bottle with a piece of cardboard and leave it for about ten minutes. Then dip into the bottle a strip of moist " ozone paper," and note that it shows the presence of ozone.

CHAPTER XXIII.

CARBON.

Occurrence.—Carbon is an example of an element which can assume three allotropic forms.

In the first it appears as a soft dull-black solid, which has no crystalline shape.[1] *Charcoal* is the most familiar example of this form of carbon.

The second variety is also soft and black, but is bright and shiny, almost like steel, and has a crystalline form. *Graphite*, or "black-lead," is the name of this variety of carbon.

The third modification is extremely hard; sometimes perfectly colourless and transparent, and highly crystalline. The name of this allotropic form of carbon is *Diamond*.

Each of these three allotropes of carbon is found in nature.

Carbon also occurs in nature in a state of chemical combination with other elements. For example, one of its compounds with oxygen, namely, the gas carbon dioxide, is present in the air, and is sent out in large volumes from rents in the rocks in volcanic districts. Again, one of its compounds with hydrogen, the gas known as marsh gas, or *fire-damp*, is found in large quantities in coal mines, and is given out by rotting vegetable matter in marshy places (hence its name, "marsh gas").

Carbon is also a constituent of all animal and vegetable matters; therefore meat, wool, bread, sugar, alcohol, wood, all contain carbon, and in many cases the carbon is associated simply with hydrogen and oxygen.

[1] Substances which show no crystalline form are called *amorphous* bodies—that is, *without form*.

Carbon is also present as a constituent of a large number of the common minerals which help to make up the solid earth. Thus, *chalk, marble, limestone, dolomite*, are all extremely common and abundant minerals, and they all contain carbon, associated with oxygen and lime, as carbonate of lime.

How to obtain Carbon from its Compounds.—When a piece of meat or bread or wood is partly burnt, we say that it is *charred.* This means that some of the hydrogen and oxygen have been expelled as water, and a portion of the carbon has been set free.

Experiment 182.—Place a little dry powdered starch in a hard glass tube, the end of which has been sealed up (see p. 37), then draw out the tube and bend it as shown in Fig. 92. This constitutes a small retort. Now heat the starch, and notice that it chars or blackens, and at the same time a liquid is expelled. This condenses in the drawn-out part of the tube and can be collected in a small test-tube. This liquid is chiefly water, resulting from the decomposition of the starch ; and the blackened mass contains free carbon. To identify the liquid as water, it will be sufficient here to drop upon it a small fragment of potassium. If the metal takes fire (as in Exp. 44) we may safely conclude that the liquid is water.

FIG. 92.

The process here illustrated is that of *destructive distillation* (see p. 179), and whenever compounds containing carbon are submitted to this treatment, carbon is set free in a more or less pure state, depending on circumstances.

Another way by which we can get carbon out of some of its compounds, is by withdrawing the other elements with which it is associated, not by fire, but by the use of some chemical reagent.

For example, sugar is a compound of carbon with oxygen and hydrogen, and we can throw the carbon out of combination by acting on the sugar with sulphuric acid.

Experiment 183.—Roughly weigh out 12 grams of loaf sugar; place it in a good-sized beaker and pour over it 10 cc. of warm water. This will dissolve the sugar in a little while, especially if gently warmed, and give a strong syrup. Now pour into this, all at once, 12 cc. of strong sulphuric acid ; the whole mixture at once froths right up. The acid takes from the sugar the hydrogen and oxygen, and leaves the carbon, which will appear as a black spongy mass.

Alcohol, or "spirits of wine," is another compound of carbon with hydrogen and oxygen, but it has a great deal more hydrogen in proportion to carbon than sugar has. Let us try and get the carbon out of some alcohol.

Experiment 184.—Pour 15 cc. of strong sulphuric acid into a little flask, and add 5 cc. of water. Cool it by dipping the flask into cold water. Then add 5 cc. of pure alcohol. [If pure alcohol is not available, use methylated spirit.] Provide a cork and delivery-tube to the flask, and arrange to collect gas in the water trough. Gently heat the flask, gas soon begins to come off. Collect first a small jar full, and allow the rest to collect in a tall cylinder until it is about one-third filled with the gas.

[If pure alcohol has been used, there will be very little blackening of the mixture in the flask, but methylated spirit contains other carbon compounds which very quickly char, or carbonize when treated in this way.]

The gas we have collected is called *ethylene*. It contains some of the carbon from the alcohol, combined with some of the hydrogen ; so that by this experiment we have not yet got *free* carbon, but have only expelled it from the alcohol still combined with hydrogen. The gas is, therefore, a *hydro-carbon*, as it is called ; that is, simply a compound of carbon and hydrogen. Hydro-carbons, as a rule, burn with a flame which gives a good light.

Experiment 185.—Remove the small jar of gas and bring a lighted taper to it. Notice the kind of flame it burns with. Contrast this flame with that of burning hydrogen.

We have learnt that chlorine has the power of taking hydrogen away from compounds of carbon and hydrogen, for in Exp. 116, when turpentine (which is also a hydro-carbon) was brought into chlorine, carbon was set free. Let us, therefore, try and get the carbon out of the gas we have obtained from alcohol, by acting on it with chlorine.

Experiment 186.—Collect a quantity of chlorine in the long cylinder containing the gas obtained in Exp. 184, until the cylinder is nearly but not quite full. Slip a glass plate over the mouth of the cylinder and shake up the remaining water in it so as to mix the gases as well as possible. Then apply a lighted taper. The mixture burns with a curious flame, producing a dense black smoke, and a black deposit of carbon all down the sides of the vessel. The chlorine has combined with the hydrogen and set the carbon free. This carbon came, therefore, originally out of the alcohol.

In these experiments it is to be noted that the carbon is always obtained in the first allotropic modification, that is, as a soft black non-crystalline substance, like charcoal or soot. It is very much more difficult to obtain carbon in the second form, as graphite ; and to make it pass into the third or diamond variety is a task which has baffled almost every attempt.

Diamond.—This form of carbon, although the most valuable in one sense, is by far the most useless so far as the chemist is concerned. It is found in gravel deposits in India, Africa, and Australia. When found, diamonds are not at all like the gems. They more often look like common little rough stones, scarcely transparent, and with little appearance of being crystals. To obtain them as usually seen in the gem, they are ground or "cut" to the desired shape, so as to "sparkle" in the light.

Diamond is the hardest known substance, and will scratch all other stones. It is, therefore, employed for cutting glass, and for giving a hard edge to drills and rock-borers. Some diamonds are brownish, and even black. These are valueless for gems, and are used for drilling and also for grinding and polishing the clear ones. When strongly heated in oxygen, the diamond first turns black and then burns, giving carbon dioxide as the only product. This proves that it is pure carbon.

Rock crystal or quartz is sometimes cut to look like diamonds. It is easy to distinguish between them by the fact that diamond burns in oxygen and gives carbon dioxide, whereas quartz does not burn at all. Quartz is not nearly so hard as diamond.

Graphite.—This form of carbon is not nearly so rare as diamond. It is found in great quantity in California. It can be made by dissolving charcoal in melted iron (just as salt is dissolved in water) and then allowing the iron to cool. As it cools carbon crystallizes out, and deposits in the form of graphite. Long before it was known that graphite was simply carbon, it was employed for writing purposes. It is so soft that when drawn across paper it leaves a black mark. Hence it got the name *plumbago*, or "the writing lead," and it is also known under the common name of *black-lead*. It must be remembered, however, that these names were given to it simply because it *looked rather like lead*. Graphite is used for making "lead" pencils, and for "black-leading" iron work. When strongly heated in oxygen, graphite, like diamond, burns, and gives the same oxide of carbon.

Amorphous Carbon.—This includes common wood-charcoal, coke, gas-carbon, lampblack, soot, animal-charcoal or boneblack. All these substances consist of non-crystalline carbon, more or less impure.

Charcoal is made by subjecting billets of wood to the process of destructive distillation (just as the starch was treated in Exp. 182) in ovens or retorts, so as to collect the volatile products as well as to secure the charcoal: or else it is made by piling the wood into stacks, and setting fire to the interior; the outside being so covered over with turf or earth as to prevent much air getting to the smouldering heap. By this method everything is lost except the charcoal. Care has to be taken to regulate the supply of air, for if too much gets into the heap the charcoal begins to burn away and is also lost.

Charcoal floats when thrown on water, but it is not really

lighter than water; it only floats because its pores are full of air, and this buoys it up.

Experiment 187.—Tie a little weight to a piece of charcoal so as to sink it, and throw it into some water in a test-tube, and then heat the water. Notice bubbles of air coming out of the charcoal. After a few minutes' boiling, allow the water and charcoal to cool, and it will be found that it will no longer float when the weight is taken off, but at once sinks in the water.

Charcoal, like both the other forms of carbon, burns in oxygen (see Exp. 59), and produces exactly the same oxide of carbon. Charcoal is very much easier to burn, however, than either diamond or graphite, and can be used to make an

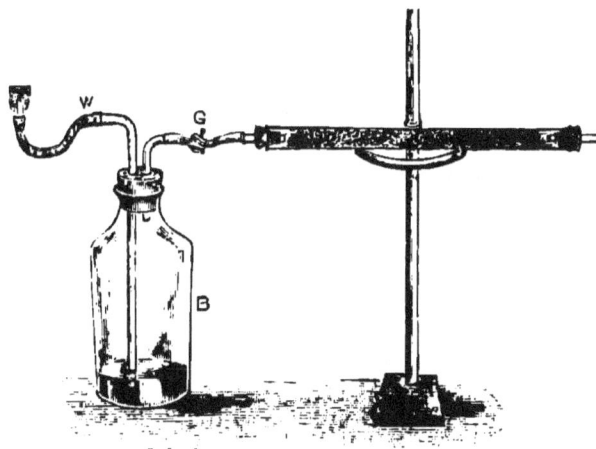

FIG. 93.

ordinary fire with. It would be quite impossible to light a fire with graphite or diamonds.

Charcoal has a wonderful power of absorbing gases, and is on this account employed to arrest the bad smelling gases arising from drains and other places.

Experiment 188.—Fit a large bottle with a cork and two glass tubes, as shown in Fig. 93, B. Break up some charcoal into little pieces and after heating them for a few minutes in a metal dish or tray, fill a piece of combustion tube with the charcoal and fit a cork and exit tube into each end.

Now place a few small particles of ferrous sulphide in a test-tube, and add a little dilute sulphuric acid. Remove the cork from the bottle B, and, by means of a piece of string, lower the test-tube for a few seconds into the bottle ; then draw it out and replace the cork. Attach the tube W to a water tap, and place a pinch-cock on G. In this way a small quantity of a gas, whose offensive smell will have been noticed, has been put into the air in the bottle B.

This gas is called *sulphuretted hydrogen*, and although its smell is quite enough to recognize it by, we can use a more convenient test.

Dip a piece of paper into a solution of lead acetate (sometimes called *sugar of lead*), and let a little of the air in the bottle blow against it, by allowing water to enter through tube W, and opening the pinch-cock on tube G.

Notice that the paper is stained black by the gas.

Now connect G to one end of the tube containing the charcoal, and gently open the pinch-cock so as to drive the air in the bottle *slowly* through the charcoal. Test the air as it passes out of the tube with another piece of paper moistened with acetate of lead. It no longer is blackened : neither will any smell of sulphuretted hydrogen be detected. The charcoal has absorbed all the bad smelling gas.

No variety of wood charcoal is quite pure carbon. For instance, if we burn a piece of charcoal there is always a certain amount of white ash left. This of course is not carbon ; but besides this, there is always a certain amount of some compounds of carbon with hydrogen.

Coke is made by treating coal very much as wood is treated in making charcoal. Coke stands very much in the same relation to coal as charcoal does to wood. Large quantities of it are produced in the manufacture of ordinary coal gas, when coal is heated in large fireclay retorts. Coke is often manufactured specially for the coke itself, either by burning coal in great stacks (like the charcoal stacks), when all the gases and liquids produced at the same time are lost ; or in special ovens or kilns, when these are caught. Coke, like charcoal, contains a certain amount of mineral matter, which is left as ash when the coke is burnt; and it also contains a little hydrogen. It is much harder and heavier than charcoal, and not so easily lighted.

Gas-Carbon is a still harder sort of coke, which is found lining the retorts in which coal-gas is manufactured. It is even harder and heavier than ordinary coke, and conducts electricity extremely well. This is the form of carbon which is employed for making the carbon rods used in electric lights.

Lampblack is the soot obtained by burning substances like petroleum or tar, which give off a large quantity of smoke. It is chiefly used for printers' ink, and for black paint.

Animal-Charcoal is the name given to the substance obtained by charring bones in iron retorts (just as wood or coal is treated). It is a very impure product, containing only about 10 parts of carbon in 100 parts, the rest being the mineral matters of the bone (chiefly phosphate of lime). When this is ground up fine it is called *boneblack*.

This substance is used chiefly for removing the dirty colour from sugar syrup, in making white sugar. It has a great power of absorbing colouring matter, and is better in this respect than any other kind of charcoal.

Experiment 189.—Half fill a wide-mouthed stoppered bottle with water, which has been tinted with either magenta, aniline blue, indigo, or some other such colouring matter. Add to this a spoonful of fine boneblack, and shake it well for a moment or two. Then filter the liquid, and note that what passes through the filter paper is quite free from colour.

All the varieties of amorphous carbon burn in oxygen and give carbon dioxide, the same product as graphite or diamond yield.

Carbon is an element which does not readily enter into chemical union with other elements. Thus, at ordinary temperatures, carbon is not acted on by hydrogen, oxygen, chlorine, or nitrogen. On this account, wood which has been charred on the surface is not so soon rotted when buried in the ground, as wood which has not been so protected. Hence it is usual to char the end of stakes or posts before putting them into the ground.

Coal is the product of a process of natural decomposition of vegetable matter which has taken long ages to complete.

The coal we burn to-day was once living vegetation. It has long been buried, owing to geological changes having taken place, and has been subjected to pressure. Coal is a very impure form of carbon, and contains compounds of carbon with hydrogen and oxygen. Roughly speaking, the proportion of carbon in soft or *bituminous* coal is about 80 in 100 parts; while in the hard or *anthracite* varieties it is about 90 parts per 100.

Although carbon is not acted on by oxygen at ordinary temperatures, it unites with it very readily when heated. Not only will it combine with free oxygen, but it will easily take oxygen away from certain oxides. On this account carbon is a most useful element to the metallurgist, for by heating oxides of metals with carbon, the carbon takes the oxygen and leaves the metal in the uncombined state.

Experiment 190.—Take a small quantity of red-lead, Pb_3O_4, and mix it thoroughly with about one quarter as much finely powdered charcoal. Put the mixture in a small porcelain crucible and strongly heat it. After a short time throw out the contents of the crucible, and there will be a globule of bright metallic lead.

This process is called *reduction*. We say that the lead has been *reduced* by the charcoal. This simply means that the oxygen with which the lead was combined has been taken from it by the carbon, and the lead, therefore, was left by itself.

This is how many of the metals are obtained from their ores. Iron ores, for instance, which are used for getting iron from, are *oxides* of iron. These when very strongly heated with carbon (usually coal or coke) are deprived of their oxygen, and the iron is set free. *All* metals cannot be reduced from their oxides in this way.

EPITOME.

Carbon is an element known in three allotropic forms. (1) Diamond. (2) Graphite. (3) Charcoal.

Diamond is the hardest known substance; highly crystalline, and often nearly or quite colourless. Used as a gem, and for cutting purposes.

Graphite is soft, black, shining, and crystalline. Used for "lead" pencils, for "black-leading."

Charcoal is soft, black, dull, and non-crystalline. Used for fuel and for making gunpowder.

Coke, lampblack, soot, bone black or animal-charcoal are all more or less impure forms of amorphous carbon.

All the varieties of carbon burn in oxygen and give the same compound, namely, carbon dioxide. In this way diamond is easily distinguished from quartz (which is silicon dioxide, and will not burn) or other imitations.

Carbon in combination occurs in all organic substances, and when these are charred, or heated so that air does not get to them to cause them to burn away altogether, the carbon is left in a more or less pure state. Wood-charcoal, animal-charcoal, and coke, are obtained by heating wood, bones, and coal in this manner.

Charcoal absorbs gases readily, and is therefore used to remove bad smelling gases ; wood-charcoal does it best.

Charcoal also absorbs colouring matter from liquids which are filtered through it or shaken up with it. Animal-charcoal does this best ; it is therefore used for filters.

Coal is an impure form of carbon, containing a number of carbon compounds.

Charcoal, coke, and coal are used as *reducing* agents, for taking oxygen away from oxides of certain metals like iron, copper, lead, etc. When such oxides are strongly heated with carbon, the metal they contain is *reduced*.

This is the principle of the *smelting* of many of the ores of metals.

CHAPTER XXIV

Carbon Dioxide, CO_2.—When a piece of charcoal is burnt in oxygen, as in Exp. 63, the carbon and oxygen combine and form the compound called carbon dioxide. Let us repeat that experiment in a slightly different way, so as to collect the compound and examine it.

Experiment 191.—Place a few little pieces of charcoal in a short piece of combustion tube, through which a stream of oxygen from

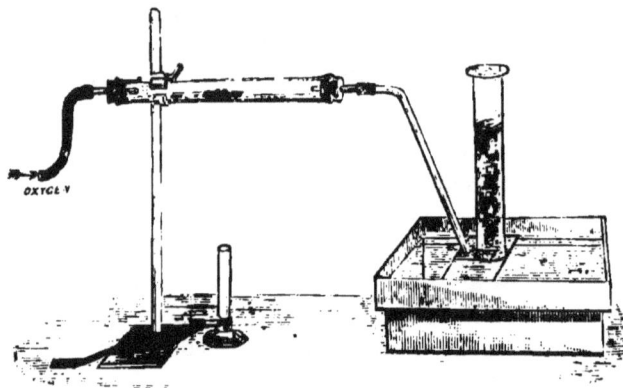

FIG. 94.

a gasholder can be passed; attach a cork and delivery tube to one end, and arrange to collect the gas over water, as shown in Fig. 94. Now heat the charcoal by means of a Bunsen flame and allow it to burn in the stream of oxygen, which should be regulated so as just to keep the charcoal burning brightly. Collect two jars of the gas.

Into one jar dip a lighted taper. Notice that the flame is instantly put out, and also that the gas does not burn. In these two respects it is like nitrogen.

Now pour into another jar some clear lime water,[1] and shake it up. Note that the lime water at once becomes milky.

This action of carbon dioxide on lime water is a test by which we can distinguish this gas from all others, as it is the only gas which can produce this effect. The equation is—

$$CaH_2O_2 \quad + \quad CO_2 \quad = \quad CaCO_3 \quad + \quad H_2O.$$

Lime water. Calcium carbonate.

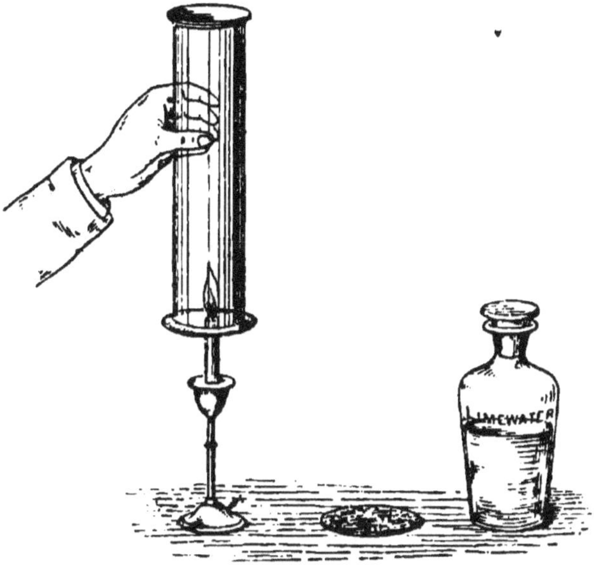

FIG. 95.

Not only is carbon dioxide produced when charcoal is burnt in oxygen, but when *any carbon compound* is burnt in the air.

Experiment 192.—Set fire to a piece of paper, and drop it while burning into a dry bottle, and cover the mouth with a piece of card or glass plate. Then pour in a little lime water, and shake it up.

Experiment 193.—Hold a dry jar over a candle flame for a minute or two, so as to catch some of the invisible products of its burning, as in Fig. 95. Then add lime water, and shake up. Do the same with a spirit lamp flame, and with an ordinary coal-gas

[1] To make lime water, place a handful of powdered lime in a big bottle, and fill it up with water, cork it up, and after shaking for a minute, leave it to settle. Then pour off the *clear* liquid into a separate bottle for use. The other bottle can be filled up with water over and over again.

flame. Notice that in every case there is abundance of carbon dioxide being produced.

Roughly speaking, every ton of coal that is burnt, produces about three tons of carbon dioxide, all of which escapes into the atmosphere.

We have already learnt that respiration is a sort of combustion, let us see if carbon dioxide is produced.

Experiment 194.—Place some lime water in a large test-tube, and by means of a glass tube make the breath bubble through the solution. Notice that the first portions of breath produce hardly any milkiness, but presently, when air that has been further into the lungs is blown through, the liquid quickly shows that there is carbon dioxide. (See also Exp. 75.)

Carbon dioxide is also given off during many processes of decay and fermentation. Thus, when vegetable matter (such as the leaves of trees which fall in autumn) gradually rots away, carbon dioxide is given off.

Experiment 195.—Dissolve a handful of sugar (common brown moist sugar is best) in some tepid water in a good-sized flask, and add some yeast. Fit a cork and delivery-tube to the flask. The sugar soon begins to ferment and gas escapes. Collect some of the gas at the water trough and test it with lime water. In breweries enormous quantities of carbon dioxide are in this way produced.

Since these different operations, namely, the burning of ordinary fuel, respiration of man and animals, decay and fermentation, besides others which are constantly going on, all result in sending carbon dioxide into the air, it is no wonder that this compound is present in the atmosphere. The wonder rather seems that the air does not become too impure to breathe. But, strange to say, if we test a small sample of ordinary air by means of lime water, we find that we can scarcely even detect the presence of this gas at all.

Experiment 196.—Take a bottle (such as has been used for testing in the above experiments) which contains simply ordinary air. Pour some lime water in and shake it up. No turbidity of the lime water is noticeable. Of course, if this experiment is made in a small room where a number of persons have been present for some time, and several gas-lamps have been burning, most likely the air will contain enough carbon dioxide to cause a turbidity in lime water when a small sample of it is tested in this way.

Since in the open air there is, after all, so small an amount of carbon dioxide (only about 3 parts in 10,000 parts of air, by measure), in spite of the enormous quantities which are every day being sent into the atmosphere—think of the millions of people in London alone, and the hundreds of tons of coal daily burnt!—it is evident that there must be some natural process at work which constantly removes this compound from the atmosphere. Nature, as usual, has made a beautiful provision for preventing the accumulation of carbon dioxide, and her secret for doing this has long been found out. She has endowed the green parts of plants with the power of decomposing carbon dioxide, by the aid of sunlight, into its two component elements, carbon and oxygen. Every green leaf and every blade of grass is a tiny chemical laboratory, where oxygen is being made from carbon dioxide, and is ever being returned to the atmosphere with all its life-supporting properties. The carbon which comes out of the carbon dioxide is utilized by the plant. Vegetation, therefore, removes dioxide from the atmosphere, and at the same time gives back the oxygen which had been taken from it by the carbon.

To prepare Carbon Dioxide for experiments, we usually adopt quite a different method.

Experiment 197.—Take the apparatus shown in Fig. 39, p. 48, and put into the bottle a quantity of broken-up marble and some water. Now pour a little strong hydrochloric acid down the funnel, and note that effervescence at once begins. After a minute or two, when the air is all swept out, collect the gas which is given off.

Marble is a variety of calcium carbonate, and the change here going on is the following—

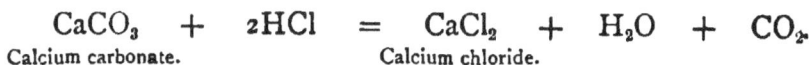

$$CaCO_3 \quad + \quad 2HCl \quad = \quad CaCl_2 \quad + \quad H_2O \quad + \quad CO_2.$$
Calcium carbonate.   Calcium chloride.

Marble is used because it is one of the most convenient of

all the carbonates; but we can equally well obtain the gas by
acting on any carbonate with almost any common acid.

Experiment 198.—Break up some sodium carbonate (common
" washing soda ") and put a little into three small beakers. Stand
each of the beakers in a larger beaker, or jar, and then by means of
a pipette, add to one a few drops of dilute sulphuric acid; to the
next, dilute hydrochloric acid ; and to the third, dilute nitric acid.
Cover the outer jars with pieces of paper. The little beakers can
now be lifted out with tongs, and lime water poured into each jar.
In each case the lime water will become turbid.

In practice we cannot use sulphuric acid with calcium
carbonate, unless the carbonate is first powdered up very fine
and made into a paste with water.

Experiment 199.—Put a few lumps of marble in a test-tube with
some water, and add a little strong sulphuric acid. Notice at first
there is an effervescence, but that it very soon leaves off, and no
more gas comes off. The first action, when it is effervescing,
is this—

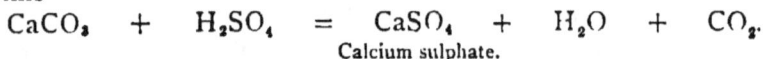

$$CaCO_3 \quad + \quad H_2SO_4 \quad = \quad CaSO_4 \quad + \quad H_2O \quad + \quad CO_2.$$
Calcium sulphate.

Calcium sulphate is formed; this is the same as " plaster
of Paris," and it at once coats over the lump of marble and
prevents any more acid getting to it.

Carbon dioxide is a very heavy gas. It is about $1\frac{1}{2}$ times
heavier than air, so that it is quite easy to collect a jar of it
by "downward displacement." More than this, the gas can
even be poured from one jar to another like a liquid; and
if poured into a vessel suspended on a balance, it will weigh
down that end of the beam.

Experiment 200.—Take a long light wooden rod, and hang from
one end of it a light cardboard box (or an old hat), and place upon
the other end a piece of bent lead rather lighter than the box.
Balance this upon the edge of a paper-knife or other convenient
metal edge, in the manner shown in Fig. 96. Then bring a good
large jar of carbon dioxide, and pour it into the box or hat, and
notice that it weighs it quite down.

Carbon dioxide is soluble to a small extent in water, but
not to such a degree as to make any difference when we

collect the gas at the water trough. All natural waters contain a *little* carbon dioxide dissolved in them, and some waters contain quite a large quantity. For instance, Seltzer water and Apollinaris water contain so much of this gas dissolved in them that they actually effervesce or "sparkle," owing to the escape of the carbon dioxide. Under ordinary conditions water dissolves about its own volume of this gas; but under increased pressure it can take up more. When such extra pressure is again removed, the extra gas that was dissolved is given off. This is illustrated in the ordinary aerated waters used for drinking. Gas is pumped into the water under great

FIG. 96.

pressure so that the water dissolves a large quantity, but the moment the pressure is released by drawing the cork, then the gas rapidly escapes with the familiar effervescence.

Experiment 201.—Colour some ordinary water with a few drops of litmus, and let carbon dioxide bubble through it for a few minutes from the apparatus used in Exp. 197. Notice that the litmus turns red, showing the presence of an acid. Note that the colour is not quite so bright red as when a drop of hydrochloric or dilute sulphuric acid is added. It is because the solution of this gas in water is acid that the gas is sometimes called "carbonic acid gas."

The acid so obtained is a very feeble acid, and easily decomposes. It is called carbonic acid, H_2CO_3. If it is gently heated it is decomposed again into water and carbon dioxide.

$$H_2CO_3 = H_2O + CO_2.$$

Experiment 202.—Take some of the solution used in the last experiment and heat it in a test-tube. Notice that the reddish colour of the litmus soon changes back again to the original blue, as the feeble acid is being decomposed.

Carbonates.—Although real carbonic acid is only a feeble acid, and so easily decomposed that it cannot be obtained by itself, it forms important salts called carbonates.

Calcium carbonate, $CaCO_3$, occurs as marble, limestone, chalk. This is the compound that is formed when lime water is brought into carbon dioxide.

Experiment 203.—Pass some carbon dioxide from the generating apparatus (Exp. 197) into some lime water until there is a thick milkiness. Filter the liquid through a small filter. The white deposit will hardly be visible on the paper, but we can test it in the following way. Put the paper into a small beaker, and pour upon it a few drops of hydrochloric acid. Notice effervescence. Put a little lime water in a wide test-tube, and pour some of the gas out of the beaker into it and shake it up.

The equation which shows what takes place when carbon dioxide is passed into lime water is—

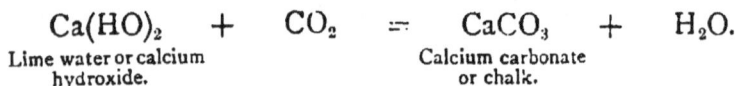

$$Ca(HO)_2 \;+\; CO_2 \;=\; CaCO_3 \;+\; H_2O.$$

Lime water or calcium hydroxide.   Calcium carbonate or chalk.

so that we get back the same compound, calcium carbonate, which was used to make the carbon dioxide from.

Experiment 204.—Pass carbon dioxide through a solution of caustic soda. Notice that there is no turbidity; also observe that the gas is eagerly absorbed, because bubbles hardly rise to the top of the liquid. Continue bubbling the gas into the solution until no more is absorbed, and then gently evaporate the solution. Take a little of the residue and add hydrochloric acid to it. Observe the effervescence. Test the gas with lime water.

Take another portion of the residue and add water to it; it dissolves easily. This is the reason why there was no precipitate formed when the gas was passed into caustic soda; the sodium carbonate that was formed is *soluble* in water. The equation in this case is—

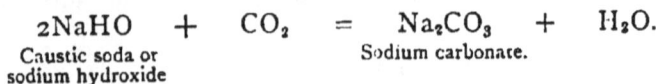

$$2NaHO \;+\; CO_2 \;=\; Na_2CO_3 \;+\; H_2O.$$

Caustic soda or sodium hydroxide   Sodium carbonate.

There are two sodium carbonates. One is common "washing soda," and the other is usually called sodium bi-carbonate. The difference in these salts is that the first contains twice as much sodium as the last; or, in other words, the last contains twice as much carbonic acid in proportion to sodium as the first. Their formulæ are—

Na_2CO_3, sodium carbonate, or normal sodium carbonate.

$HNaCO_3$, sodium bi-carbonate, or hydrogen sodium carbonate.

Experiment 205.—Dilute some lime water with about half as much distilled water, and pass carbon dioxide through it. As in the former experiment there is a precipitate of calcium carbonate.

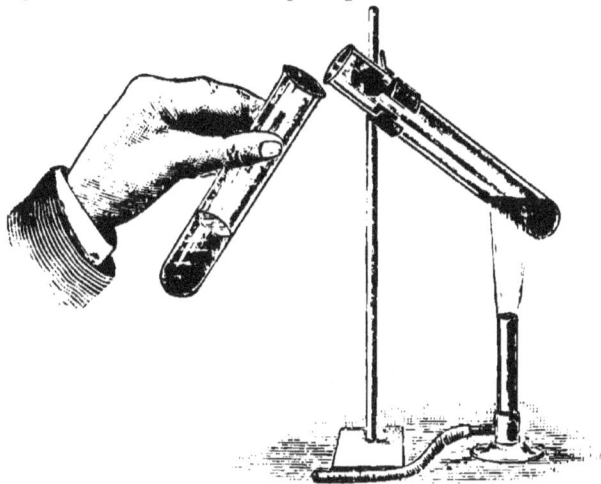

FIG. 97.

But continue passing the gas, stirring up the liquid, and notice that in a few minutes it becomes perfectly clear again. The calcium carbonate has dissolved in the solution of carbon dioxide. Take a little of this clear solution and boil it. It again becomes turbid, because the carbon dioxide is expelled, and the calcium carbonate cannot remain in solution. The presence of calcium carbonate dissolved in this way is what causes the temporary hardness of natural waters (see pp. 82 and 269).

Many carbonates part with carbon dioxide when they are heated, leaving an oxide.

Experiment 206.—Heat a little magnesium carbonate in a test-tube, and allow the gas to flow down into another test-tube, containing lime water, in the manner indicated in Fig. 97,

$$MgCO_3 = CO_2 + MgO.$$

Calcium carbonate undergoes the same change, but requires a higher temperature.

The process of lime burning, carried on in lime kilns, illustrates this.

Limestone (that is, calcium carbonate) is heated in the kiln, when carbon dioxide escapes into the air, and lime (that is, calcium oxide) remains.

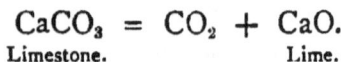

$$CaCO_3 = CO_2 + CaO.$$

Limestone.			Lime.

The " setting " of mortar is partly due to the absorption of carbon dioxide out of the air by the lime in the mortar. Old mortar, therefore, contains calcium carbonate.

Experiment 207.—Pour a little hydrochloric acid upon some fragments of old mortar · notice the effervescence, and pass the gas into lime water.

To find the Weight of Carbon Dioxide in Marble.—

Experiment 208.—Select a small thin flask with a wide mouth, and fit it with a cork having two holes. Through one hole insert a short straight glass tube, the top of which can be closed with a tiny cork. Into the other hole fit a bent glass tube with a small bulb blown on it, as shown in Fig. 98. Make a hole in the bottom of a small test-tube, by heating the end with a fine-pointed blow-pipe flame, and blowing down the test-tube. Then cut off the tube about $3\frac{1}{2}$ cm. ($1\frac{1}{2}$ inches) from the bottom, and "border" the end (p. 36) so as to obtain the little apparatus shown at A, Fig. 98. Fasten a thin copper wire to the neck, and hang the little tube inside the flask in the manner shown. Place a little strong sulphuric acid in the bulb tube B, so as to just fill the bend, and close the open end with a little cap, C

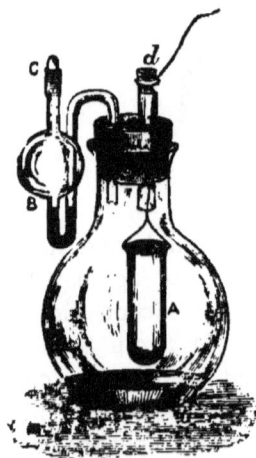

FIG. 98.

(p. 111). Put about 10 or 12 cc. of a mixture of equal parts of hydrochloric acid and water in the flask, but do not let it touch the little hanging tube. Now carefully weigh the whole apparatus. Then remove the cork and drop into the hanging tube one or two little fragments of marble (*not so small as to drop through the hole in the bottom*), and weigh again. The increase gives the weight of marble, which should be from 1 to 2 grams.

Remove the cap C, and then lower the tube containing the marble so that it dips into the hydrochloric acid, and replace the little cork *d*.

The acid dissolves the marble, and the carbon dioxide which is evolved makes its escape by bubbling through the sulphuric acid in B. This prevents the gas from carrying away any vapour of water with it.

When the marble has all dissolved, the apparatus of course is full of carbon dioxide, and as this is much heavier than air it must be removed. Take out cork *d*, and by means of a short piece of rubber tube attached to the bulb-tube, slowly suck the gas out of the flask, *making it bubble quite slowly* through the sulphuric acid in the tube. When the gas is all out, and the apparatus is full of air as it was at first, replace the cork *d*, and cap C, and weigh again. The loss of weight represents the carbon dioxide which has passed out. Two experiments should be made, so as to confirm the result.

EXAMPLE.   Weight of marble used = 1·5 grams
Weight of carbon dioxide = 0·6   „
as 1·5 : 100 :: 0·6 = 40

therefore this sample of marble contains 40 per cent. of carbon dioxide.

EPITOME.

Carbon dioxide is produced when all carbon compounds burn. Coal, gas, candles, oil, wood, etc., when burnt give carbon dioxide amongst other things. The breathing of animals and man produces carbon dioxide, and it is given off during processes of decay and fermentation.

It is also formed when limestone (that is, calcium carbonate) is heated, as in the process of lime-making in lime kilns.

Carbon dioxide is prepared by acting on calcium carbonate (marble, chalk, or limestone) with hydrochloric acid.

Carbon dioxide is a colourless gas, slightly soluble in water forming a feebly acid solution, which is decomposed again when heated. The gas does not burn, and puts out ordinary flames. Carbon dioxide is not exactly poisonous, because we are always breathing small quantities of it in the air, but if animals are placed in the gas they quickly die for want of oxygen. Even a moderate

---

[1] The student should calculate what weight of carbon dioxide would have been obtained if the marble had been *pure* calcium carbonate.

quantity of this gas in the air, over and above the usual amount, is injurious to life.

Carbon dioxide is a heavy gas, which can be collected by downward displacement, and can be poured from one vessel to another.

When passed into lime water it unites with the lime and forms a white precipitate of chalk.   It is quickly absorbed by either caustic soda or potash, forming sodium or potassium carbonates.

The test for a carbonate is to add an acid to it, and to prove that the gas which is given off is carbon dioxide by passing it through lime water and obtaining the precipitate of chalk.

CHAPTER XXV

Carbon Monoxide, CO.—It has been shown that when carbon burns in air or oxygen under ordinary circumstances it gives carbon dioxide, but under particular conditions the other oxide is formed. For instance, if we fill a long piece of combustion tube with fragments of charcoal, and heat the tube to a good red heat in a furnace, and then pass a moderate stream of oxygen in at one end, instead of getting carbon dioxide coming out at the other, we should find that it was quite a different gas; for we could set fire to it, and should see it burn with a beautiful blue flame. What takes place is this. As the oxygen first meets the hot charcoal, the charcoal burns and greedily takes all the oxygen it can get, and produces carbon dioxide. But as this passes along the red hot tube it gives up half of its oxygen to a further portion of carbon, so that they go equal shares, as it were, with the oxygen, and the result is carbon monoxide.

$$CO_2 + C = 2CO.$$

We can prove that this is the true explanation in the following way.

Experiment 209.—Take a piece of ordinary iron gas-pipe and fill it with fragments of charcoal and make it red hot in a furnace (Fig. 99). (If more convenient, the iron pipe may be bent in the middle into a sort of elbow shape, and the bent part pushed into an ordinary fire, between the bars.)

Now we must not pass oxygen direct into this pipe, or the iron itself would burn up (refer back to oxygen); so we must connect to one end of it a short piece of glass combustion tube containing a few pieces of charcoal. Now heat the charcoal in the glass tube and pass a gentle stream of oxygen through. The carbon burns, and the carbon dioxide passes along over the red hot charcoal in the iron pipe. Set fire to the gas which escapes at the other end, and observe its flame. This gas is carbon monoxide.

Carbon monoxide is always being produced by exactly this method in our ordinary fireplaces. The oxygen of the air which enters the front and bottom of the grate is taken up by the first portion of burning coal, and forms with the carbon carbon dioxide. This, in passing through the fire, meets with red-hot carbon, and gives up half its oxygen, forming carbon

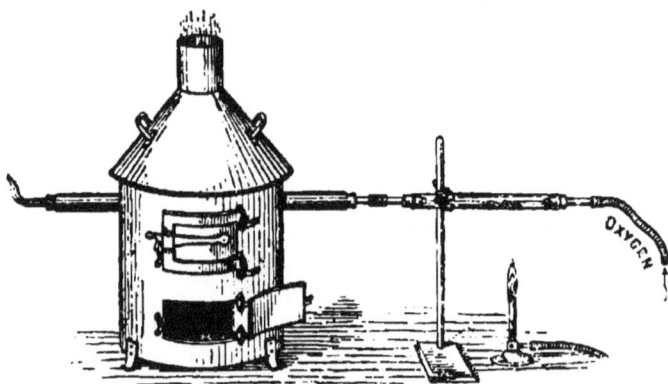

FIG. 99.

monoxide. This operation always goes on in the fire, but to a greater extent if the fire is hot and clear, and especially in a coke fire.

A good deal of the carbon monoxide thus formed burns on the top of the fire, and the familiar bluish flame seen flickering about on the top of a clear bright fire is this gas burning. But some of it escapes unburnt up the chimney, because there is often so much carbon dioxide passing off from the fire as to prevent the other gas from burning.

If instead of passing carbon dioxide through the tube of red-hot charcoal, we were to send steam through it, we should then get a mixture of carbon monoxide and hydrogen,

$$C + H_2O = CO + H_2.$$

This mixture is sometimes called *water-gas*, and is made

on a large scale by passing steam over strongly heated coal or coke.

Preparation of Carbon Monoxide.

Experiment 210.—Put a few crystals of oxalic acid into a test-tube, and pour upon them a little strong sulphuric acid. Fit a cork and delivery tube to the test-tube and gently heat the mixture. Notice that a brisk effervescence soon begins to take place. Collect two jars of the gas, over water. Test the gas in one jar with a lighted taper, and note that it burns with a blue flame, but does not seem to burn very well.

Into the other jar pour some lime water ; notice that milkiness is at once produced. This shows that *carbon dioxide* is also present, because no other gas produces this turbidity with lime water.

To find out how much carbon dioxide is present, we can make use of the fact that this gas is quickly absorbed by caustic soda.

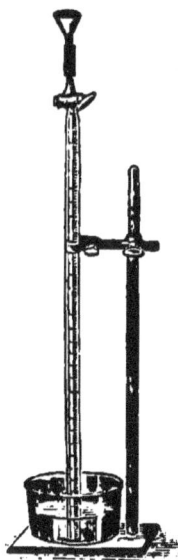

Experiment 211.—Collect some more of the gas given off by the oxalic acid in a burette (fill the burette with water, and invert it in a basin or trough, just as an ordinary gas-collecting jar). When it is full of gas, attach a small funnel to the top with a piece of rubber tube, as shown in Fig. 100. Pour some caustic soda solution into the funnel, and then gradually turn the tap so as to allow the liquid slowly to enter the tube and trickle down the side. Close the tap before the funnel is empty, or else air will be drawn in. Notice that as the caustic soda enters, the gas is

FIG. 100.

quickly absorbed by it, and the water rises in the tube. When it stops rising, note how much gas remains ; almost exactly one half.

Close the tube with the thumb, remove it from the trough and invert it. Then bring a lighted taper to the gas. Notice that it burns with the blue flame of carbon monoxide.

This shows that, when oxalic acid is acted on by sulphuric acid, carbon dioxide and monoxide are given off *in equal volumes.* The equation expressing the change is this—

$$C_2H_2O_4 = CO_2 + CO + H_2O.$$
Oxalic acid.

The water is taken up by the sulphuric acid, which has a powerful affinity for water.

When we want carbon monoxide pure, we must pass the gas obtained from oxalic acid through bottles containing caustic soda, so as to get the carbon dioxide removed.

Experiment 212.—Fit up the apparatus shown in Fig. 101. Place in each of the two bottles some solution of caustic soda, and heat the mixture of oxalic acid and sulphuric acid in the small flask. As the gases bubble through the caustic soda, the carbon dioxide is absorbed, and the monoxide passes on. Collect three jars.

FIG. 101.

Dip a lighted taper into one. Notice that the gas burns with a much stronger flame than before. Also note that the taper itself is extinguished if thrust into the gas.

Add lime water to the second. Is there any turbidity produced? If so, it proves that even bubbling through two bottles has not quite removed all the carbon dioxide. Take the third jar, and quickly pour into it a little caustic soda and cover it again immediately. Shake the liquid up with the gas, and then replace the jar in the trough for a few minutes, in order to let the caustic soda go out into the water. Now take the jar out again, and pour some lime water into it and shake up. This time there should be no turbidity at all. Now light the gas, and as the flame passes down into the jar, cover it with the glass plate. Again shake up the lime water that is in the jar, and notice that now it is instantly made milky.

This shows that when carbon monoxide burns, it gives carbon dioxide.

$$CO + O = CO_2.$$

Carbon monoxide takes oxygen away from many metallic oxides when they are strongly heated in this gas. It is, therefore, like carbon, a reducing agent. Thus, if carbon monoxide is passed over heated oxide of iron, the oxide is deprived of its oxygen and becomes reduced to the metallic state, and carbon dioxide is formed.

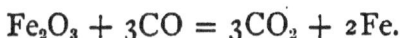

$$Fe_2O_3 + 3CO = 3CO_2 + 2Fe.$$

This process goes on in the blast furnace where iron ores are smelted.

Carbon monoxide is extremely poisonous. Many people have been killed by the gas escaping from coke or charcoal fires, their deaths being due to the poisonous nature of the carbon monoxide which such fires give off.

EPITOME.

Carbon monoxide is produced when carbon burns in an insufficient supply of oxygen, or when carbon dioxide is passed over red hot charcoal.

It is prepared from oxalic acid by the action of sulphuric acid, the gas so obtained being passed through caustic soda to remove the carbon dioxide. Carbon monoxide is a colourless poisonous gas, which burns with a beautiful blue flame. When it burns it produces carbon dioxide.

Carbon monoxide does not dissolve in water ; does not make lime water turbid ; is not absorbed by caustic potash. It is not an acid forming oxide like carbon dioxide, and, therefore, forms no salts.

It is readily distinguished from all other gases by burning with a blue flame, and forming carbon dioxide.

CHAPTER XXVI.

SULPHUR.

THIS element is found chiefly in volcanic regions, such as Sicily and Iceland, where it exists in the free state; that is, not in chemical combination with other elements. The sulphur as thus found is called *native sulphur*.

Besides occurring in this uncombined state, sulphur is also a constituent of a large number of important ores, in which it is combined with various metals. Some of the commonest of these sulphides are *iron pyrites* (sulphur combined with iron, FeS_2); *copper pyrites* (sulphur with copper and iron, $CuFeS_2$); *zinc blende* (sulphur and zinc, ZnS); *galena* (sulphur and lead, PbS).

In combination with metals and with oxygen together, it is found in *heavy spar* (barium sulphate, $BaSO_4$); and the very common mineral *gypsum* (calcium sulphate, $CaSO_4 + H_2O$).

Modes of obtaining Sulphur.—(1) "Native sulphur" is always mixed up with more or less earthy and mineral matters. In order to separate the sulphur, the crude material is piled up into heaps on a slanting hearth, and the heaps set on fire. Some of the sulphur burns away, but the heat it gives out melts the remainder, which runs away from the impurities down the sloping floor. Of course the supply of air to the heap requires to be regulated, for if it had free access, the whole of the sulphur would be burnt away. (2) Sulphur can also be got from iron pyrites, by heating the ore strongly, without letting air get to it.

Experiment 213. Heat a little powdered iron pyrites in a test-tube. Notice the sublimate which collects on the cooler part of the tube, as it forms little drops of melted sulphur. After a time, just touch the hot end of the test-tube with a drop of water so as to crack it off. Now heat the sulphur on the side of the tube, holding the tube in an inclined position. Notice the sublimed sulphur takes fire and burns, and the gas which escapes at the top of the tube has the characteristic choking smell produced by burning sulphur.

The pyrites does not part with *all* its sulphur when heated in this way. The change is expressed by the equation—

$$3FeS_2 = Fe_3S_4 + 2S.$$

If the iron pyrites is roasted in a free current of air, it loses *all* its sulphur, and both the iron and the sulphur are converted into oxides, thus—

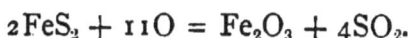

$$2FeS_2 + 11O = Fe_2O_3 + 4SO_2.$$

A large quantity of sulphur is now recovered from waste products which contain this element, and which used to be thrown away. Such a refuse substance as that known as *Alkali-waste* (obtained in the process of manufacturing sodium carbonate) is now utilized in this way.

Purification of Sulphur.

Experiment 214.—Bend a piece of wide glass tube, one end of which is closed up, and attach it to a retort in the manner shown in Fig. 102. Place a few pieces of sulphur in the little bent tube, and apply heat to it. The sulphur melts ; then gets very dark in colour, and presently boils, and the vapour passes into the body of the retort. The sulphur is here being distilled, and the impurities it may contain remain behind. Notice that,

FIG. 102.

as the vapour enters the large retort, some of it condenses upon the glass as a yellowish powder, while some collects on the lower part of the vessel in the liquid state and quickly solidifies.

On a large scale this process is carried out by boiling the sulphur in earthenware retorts, and sending the vapour into large brickwork chambers. At first it condenses as a fine light yellow powder on the walls. In this condition it is called *flowers of sulphur*. After a time the walls of the chamber get warm, and the sulphur melts and collects on the floor, and is then run out into wooden moulds so as to cast it in the form of sticks. This is called *roll sulphur*, or brimstone.

Properties of Sulphur.

Experiment 215.—Take a piece of common roll sulphur, and strike it gently on the table, or with a pestle. Notice how brittle it is. Examine the freshly broken surfaces and see that it is highly crystalline. Powder a little of it in a mortar, and try if it dissolves in water. After shaking it up with water in a test-tube for a little time, decant some of the water into a small dish and evaporate gently to dryness. If there is nothing left it shows that sulphur is not dissolved by water.

When sulphur is heated it behaves in a rather striking manner.

Experiment 216.—Carefully heat a little of the powdered sulphur in a test-tube; it easily melts, and, if not over-heated, gives a pale amber-coloured liquid which runs about the tube like oil. Now heat more strongly, and notice that the sulphur rapidly deepens in colour, becoming like dark treacle, and gets so thick and sticky that if the tube is turned upside down it does not run out at all. Heat it still more, and note that it becomes quite liquid again, although remaining dark coloured, and presently boils. It is difficult to see the colour of the vapour, because of the almost black appearance of the liquid; but by looking through the tube at any part where the glass is clear, it will be seen that the vapour has a *pale yellow colour*. Let the test-tube cool, and the sulphur goes through the same changes in the opposite order. Notice that as it solidifies it forms crystals on the sides of the tube.

Sulphur melts at $114 \cdot 5°$; at a temperature about $230°$ it passes into the thick condition, and at $448°$ it boils.

Allotropic Modifications.—Sulphur, like carbon, exists in three allotropic forms; and, like those of carbon, two are crystalline, and one non-crystalline or amorphous. But here the similarity ends.

Sulphur is quite easily made to assume either of its three forms (not so, carbon), but it will only *remain* in one of them, for both the other varieties gradually pass back into the first. (Carbon is stable in each of its allotropic forms.)

Experiment 217.—Put some fragments of roll sulphur into a test-tube, and add to them a small quantity of a liquid called carbon disulphide (CS_2) just to cover them. Notice that the sulphur quickly dissolves. Pour the solution into a small dish, cover it with a piece of cardboard, and leave it for some time to slowly evaporate. Examine the residue with a pocket lens, and carefully note the shapes of the crystals.

If this experiment is made on a larger scale, and with certain precautions, more perfectly shaped crystals will be obtained, like the one shown in Fig. 103.

The form of the crystals of sulphur obtained in this way is what is known as *Rhombic Octahedral.* They have a beautiful amber-like appearance, and are very brittle.

The sulphur which is found "native" is in this form.

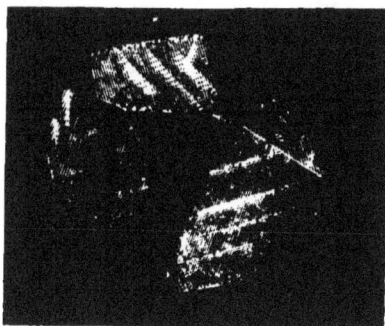

FIG. 103.

Experiment 218.—Carefully melt some roll sulphur in a small beaker; the beaker being about three-quarters full. Allow it to cool, carefully watching it, and as soon as a thin crust has formed on the top, pour out what remains of the liquid into a dish or plate. On cutting away the crust, the interior of the beaker will be found to be lined with long needle-shaped crystals, like those shown in Fig. 104. Examine these crystals, and note how entirely different they are in shape to the others. They have also a transparent appearance.

FIG. 104.

This is the second allotropic form of sulphur, and is known by the name *Prismatic Sulphur*, because the crystals are in the shape of long thin prisms. If these crystals are kept for a day or two they lose their transparent appearance, and become exactly the colour of ordinary brimstone. In fact, they change back again from the *prismatic* modification to the *rhombic* form; and although the crystals retain the outward form of the prism, they crumble on the slightest touch to a number of minute crystals, having the shape of rhombic octahedrons.

Experiment 219.—Heat a quantity of sulphur (either "flowers" or powdered lump) in a common oil flask until it gets into a boiling

FIG. 105.

condition. Then pour the hot sulphur in a thin stream into cold water in a beaker. (A funnel may be stood in the water, and the stream of sulphur poured round and round it, Fig. 105.) Now lift out the funnel with the congealed sulphur, and notice how entirely different it is from either of the other forms. It is no longer brittle, but seems almost like indiarubber, and can be stretched and pulled into threads.

This is the third allotrope of sulphur, and is called *Plastic Sulphur*. It has no crystalline character at all.

Like the prismatic form, this plastic sulphur when left to itself for a few days changes back again into the rhombic variety. It gradually loses its curious elasticity, and becomes the ordinary brittle, yellow, crystalline sulphur. If it is stretched about, or slightly warmed, it changes from its plastic state to the ordinary condition much more quickly.

Milk of Sulphur is an old-fashioned name for sulphur obtained in the following way.

Experiment 220.—Throw a small handful of flowers of sulphur into a saucepan half full of hot water, and add about twice as much lime. Boil the mixture for five or ten minutes, and then allow it to settle. Take some of the clear yellowish liquid and add to it a little strong hydrochloric acid. A white precipitate of sulphur in the condition of very fine particles is produced, which makes the liquid look almost like milk ; hence the name " Milk of Sulphur."

This is a much finer powder than flowers of sulphur, and is on this account used in medicine.

When sulphur, in any of its allotropic forms, is heated in air, it burns with a blue flame (see Exps. 64, 138), and gives an oxide of sulphur called sulphur dioxide, SO_2.

Combination of Sulphur with Metals.—Sulphur combines with many metals, and forms compounds called *sulphides ;* just as oxygen unites with metals and gives *oxides*.

Experiment 221.—Heat a small quantity of flowers of sulphur in a test-tube until it boils, and drop in a fragment of sodium about the size of a large pin's head. The sodium instantly takes fire and burns brilliantly, producing sodium sulphide, Na_2S.

Experiment 222.—Drop into a similar small quantity of boiling sulphur some reduced iron (that is, iron obtained by heating iron oxide in a stream of hydrogen or coal gas). The black powder at once takes fire as it comes into the sulphur vapour, and gives ferrous sulphide, FeS. If iron filings are used instead of the "reduced" iron, they combine with the sulphur, but being so much coarser they do not take fire ; but if a fine iron wire is made red hot in a Bunsen flame, and then plunged into the sulphur vapour, the wire burns, and melted drops of iron sulphide fall to the bottom of the tube.

Thin copper wire also readily combines with sulphur vapour, and takes fire without being heated in a lamp, producing copper sulphide, CuS.

EPITOME.

Sulphur occurs in volcanic regions, such as Sicily and Iceland, in the free or elementary state. In combination with many metals, in some of the commonest ores of these metals, as *sulphides ;* also in combination with metals and oxygen, as *sulphates*.

" Native sulphur " is burnt in heaps, with a limited air supply, so as to melt the sulphur and separate it from rocky matter with which it is mixed.

Sulphur is got from iron pyrites by simply heating it, when it gives off one-third of its sulphur.

Sulphur is purified by distillation ; and is obtained either as *flowers*, or is melted and cast into sticks, known as *roll* sulphur.

R

The three allotropic forms of sulphur are—

(1) *Rhombic* (octahedral). Permanent, soluble in carbon disulphide. Specific gravity, 2·05. Lemon yellow colour, very brittle.

(2) *Prismatic.* Not permanent, slowly changes to No. 1. Translucent amber yellow crystals. Specific gravity, 1·98.

(3) *Plastic.* Not permanent, slowly changes to No. 1. Soft, non-crystalline, indiarubber-like, translucent yellow. Not dissolved by carbon disulphide. Specific gravity, 1·95.

Sulphur (any variety) burns in the air with a blue flame, producing sulphur dioxide.

Sulphur is not soluble in water, but is oxidized by nitric acid into sulphuric acid (see p. 191). It unites directly with metals, forming sulphides.

Sulphur belongs to a family of elements, the members of which are oxygen, sulphur, selenium, and tellurium.

The last two are rather rare elements.

The chemical relationship between sulphur and oxygen will be seen by comparing some of the compounds which each forms with other elements.

Water, H_2O	Sulphuretted hydrogen, H_2S.
Potassium hydroxide, KHO	Potassium hydrosulphide, KHS.
Calcium hydroxide, $Ca(HO)_2$	Calcium hydrosulphide, $Ca(HS)_2$.
Carbon dioxide, CO_2	Carbon disulphide, CS_2.
Potassium, oxide, K_2O	Potassium sulphide, K_2S.
Copper oxide, CuO	Copper sulphide, CuS.

Occurrence.—This compound is a gas, and is present in volcanic gases. It is met with, dissolved in water, in certain sulphur springs, such as those at Harrowgate. When animal substances containing sulphur become putrid, sulphuretted hydrogen is formed; bad eggs owe their disagreeable smell to the presence of this gas. Ordinary coal-gas, as it leaves the retorts, contains a considerable amount of sulphuretted hydrogen. This is removed in the " purifiers " before the gas is sent out from the gasworks, and the sulphur it contains is extracted and sold.

Modes of Formation.—Sulphur does not very easily combine with hydrogen ; but if hydrogen is passed over boiling sulphur, a portion of the sulphur unites with hydrogen and forms sulphuretted hydrogen.

Experiment 223.—Heat a few fragments of sulphur in a horizontal bulb tube, until the sulphur boils, and then allow hydrogen to pass through the tube. Smell the gas escaping at the end. Also test it by holding a piece of paper moistened with a solution of lead acetate in the gas. The paper becomes black, showing the presence of sulphuretted hydrogen.

In the laboratory, the gas is always prepared by another method, namely, by acting on ferrous sulphide with either sulphuric or hydrochloric acid. Ferrous sulphide is made by heating iron and sulphur together. Its composition is expressed by the formula FeS. This substance must not be confounded with iron pyrites, FeS_2.

Experiment 224.—Place a quantity of ferrous sulphide in a two-necked bottle (see Fig. 39), cover it with water, and pour a small quantity of strong sulphuric acid through the thistle funnel. Notice that effervescence immediately begins. After a few minutes, during which the air in the bottle is being gradually swept out, collect two jars of the gas, using water in the trough as warm as the hands can comfortably bear.

The reaction in this case is—

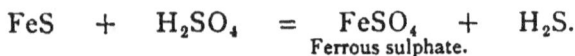

$$FeS \; + \; H_2SO_4 \; = \; \underset{\text{Ferrous sulphate.}}{FeSO_4} \; + \; H_2S.$$

If hydrochloric acid is used instead of sulphuric acid the equation is—

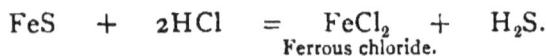

$$FeS \; + \; 2HCl \; = \; \underset{\text{Ferrous chloride.}}{FeCl_2} \; + \; H_2S.$$

Then remove the delivery tube from the apparatus and light the gas as it escapes from the exit tube. Notice that the flame is bluer than that of hydrogen, but not so blue as that of burning sulphur. Gently smell the products of the burning gas ; note that there is the same choking smell of sulphur dioxide, as when sulphur burns. Hold a cold tumbler over the flame, and observe the moisture collecting.

When sulphuretted hydrogen burns from a jet with a free supply of air, it gives sulphur dioxide and water–

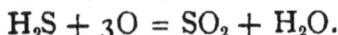

$$H_2S + 3O = SO_2 + H_2O.$$

Experiment 225.—Depress a piece of glass down on to the flame, and note that there is a deposit of sulphur.

Experiment 226.—Test the gas in one of the jars with a lighted taper, note that the taper is extinguished when thrust into the gas. Observe also that, as the gas burns, there is a deposit of sulphur formed on the sides of the jar. This is from the same cause as in the last experiment, namely, because the supply of air is not sufficient to completely burn the gas. We may express it by this equation—

$$H_2S + O = H_2O + S.$$

Experiment 227.—Transfer the second jar of gas to a trough of cold water, and leave it standing mouth downwards. Notice that the water rises a little in the jar. Shake it up as much as possible without lifting the mouth out of the water, and in a short time the water will have absorbed nearly all the gas. Smell the water ; note that it smells like the gas.

Sulphuretted hydrogen is considerably soluble in cold water. At the common temperature, water dissolves about three times its own bulk of this gas. The solution is called sulphuretted hydrogen water. Warm water dissolves much less of the gas, therefore we usually collect it over hot water, as described in Exp. 224.

Sulphuretted hydrogen made from ferrous sulphide is always mixed with free hydrogen, because the ferrous sulphide always contains some iron which is not combined with sulphur ; and when the acid comes in contact with this free iron, hydrogen is evolved (see Hydrogen, p. 46). Therefore, if sulphuretted hydrogen is required quite pure, we generally use antimony sulphide, Sb_2S_3. For the purposes for which the gas is generally required, however, the presence of a little hydrogen does not matter, so that in practice it is nearly always made from the iron compound. For a number of experiments we want just a few bubbles of sulphuretted hydrogen, and therefore it is convenient to have an apparatus so arranged for making the gas that we can stop the action when we please, and start it again. This cannot be done with the form of apparatus used for Exp. 224, without emptying the bottle each time. A very simple form of constant apparatus can be made in the following way.

Experiment 228.—Obtain two large test-tubes (" boiling tubes ") and draw them out at one end, as shown in Fig. 106. Secure one of them with wire or thread to a retort stand, and join their drawn-out ends with a piece of indiarubber pipe in the manner shown. Half fill the fixed one with small broken pieces of ferrous sulphide, and close the tube with a cork and exit tube, the latter carrying a short piece of rubber tube, *t*, with a screw clamp, *s*, upon it.

Suspend the other boiling tube to a ring by means of a wire hook. Close the clamp, and pour dilute sulphuric acid into the open tube until it is about three-quarters full. Now gently open the clamp, when the acid will gradually enter the other tube, and, coming in contact with the ferrous sulphide, will cause the evolution of sulphuretted hydrogen. So long as the clamp is open, the gas will escape from the tube, but as soon as it is closed again, the gas, which is still being produced, not being able to get out, begins to drive the acid back into the open tube, and then, of course, the action stops. In this way, by opening and closing the clamp, we can produce a little gas and then stop the supply at will.

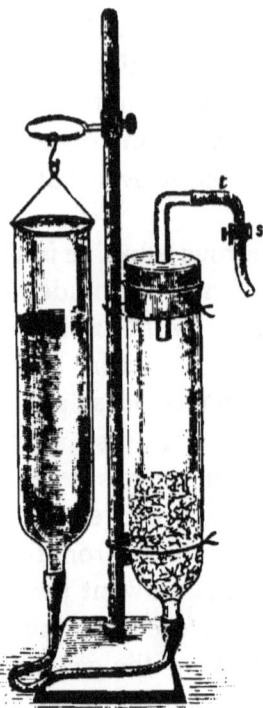

FIG. 106.

By means of this little apparatus for getting sulphuretted hydrogen, do the following experiments—

Action of Sulphuretted Hydrogen on Metals.

Experiment 229.—Open the screw clump, *s*, and let the gas blow against a clean silver coin for a moment. Notice that the silver is at once turned black. [A silver spoon is stained in the same way by a bad egg.] The black substance is silver sulphide, Ag_2S.

Experiment 230.—Place a small fragment of potassium in a horizontal bulb tube, and pass sulphuretted hydrogen through the tube. Heat the potassium, when it will burn brightly in the gas. Sulphuretted hydrogen, therefore, will support the combustion of potassium. The reaction is—

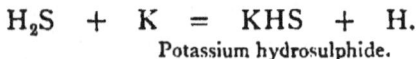

$$H_2S + K = KHS + H.$$
Potassium hydrosulphide.

Compare this with the action of potassium upon water—

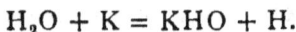

$$H_2O + K = KHO + H.$$

Action of Sulphuretted Hydrogen on Metallic Compounds.

Experiment 231.—Let sulphuretted hydrogen blow against a piece of ordinary canvas, used by artists. Notice that it is quickly blackened. The canvas is painted with white lead (a lead carbonate), this is at once converted into lead sulphide, which is black.

[There is always a little sulphuretted hydrogen in the air of towns, and this makes oil paintings gradually turn black.]

Experiment 232.—Put a little oxide of iron in a horizontal glass tube, and gently warm it. Then pass sulphuretted hydrogen through the tube. Notice that the reddish oxide is turned black, and glows with the heat produced by its combination with the sulphur—

$$Fe_2O_3 + 3H_2S = 2FeS + S + 3H_2O.$$

Similarly if sulphuretted hydrogen is passed over slaked lime (calcium hydroxide), we get a sulphur compound of calcium produced.

$$\underset{\text{Calcium hydroxide.}}{CaH_2O_2} + 2H_2S = 2H_2O + \underset{\text{Calcium hydrosulphide.}}{CaH_2S_2.}$$

These two substances, namely, slaked lime and iron oxide, are the materials used in the " purifiers " of the gas-works, to absorb the sulphuretted hydrogen from the coal-gas.

Action of Sulphuretted Hydrogen on Metallic Solutions.

Experiment 233.— Prepare the three following solutions :— (1) Dissolve a small particle of lead acetate in water in a test-tube. (2) Dissolve a similar quantity of white arsenic (arsenious oxide) in a few drops of hydrochloric acid, and add water to half fill the test-tube. (3) Dissolve a small quantity of *tartar emetic* (an antimony salt) in water.

Now pass a few bubbles of sulphuretted hydrogen through each of these, by dipping a tube from the apparatus (Fig. 106) into the solutions in turn.

Notice what happens in each case. We get a black, a yellow, and a red precipitate. In other words, the three elements, lead, arsenic, and antimony, form sulphides ; lead sulphide being black, arsenic sulphide yellow, and antimony sulphide red.

Sulphuretted hydrogen, therefore, affords a test by which we can easily distinguish between compounds of these elements.

Experiment 234.—In three separate test-tubes put a little dilute solutions of (1) copper sulphate, (2) ferrous sulphate, and (3) potassium nitrate ; add to each, one or two drops of hydrochloric acid, and pass into each a few bubbles of sulphuretted hydrogen. Note that there is a precipitate only in the copper solution.

This means that copper sulphide is precipitated in an *acid* solution ; whereas no sulphide of either iron or potassium is produced in a solution containing hydrochloric acid.

$$\text{In acid solution, } CuSO_4 + H_2S = CuS + H_2SO_4.$$

Now add a few drops of ammonia to the two solutions which were not precipitated, and note that in the case of the iron salt, a black precipitate is produced, but nothing happens in the solution of the potassium salt.

In alkaline solution, $FeSO_4 + H_2S = FeS + H_2SO_4.$

This black precipitate is ferrous sulphide ; and this experiment shows that in an *alkaline* solution iron is precipitated in the form of sulphide, while no potassium sulphide is produced either in the acid or alkaline solutions.

The following experiment shows what use we can make of these facts—

Experiment 235.—Pour a little of the solutions of copper sulphate, ferrous sulphate, and potassium nitrate into the same test-tube. Add a drop or two of hydrochloric acid, and bubble sulphuretted hydrogen through the solution for a few minutes. We know by the former experiment that only the copper will be precipitated as sulphide.

Now filter the liquid. The copper sulphide remains on the filter, while the solution which passes through contains the iron and potassium salts. Next add to the filtrate some ammonia. This we have seen causes the precipitation of ferrous sulphide (there being sulphuretted hydrogen dissolved in the solution). Pass this through another filter. The ferrous sulphide remains on the filter and the potassium salt passes through.

In this way we have *separated the three metals, copper, iron, and potassium,* whose salts were originally mixed in the solution.

Sulphuretted hydrogen is, therefore, a most important agent in analysis ; for we find (1) a certain number of metals which are precipitated as sulphides in acid solution, (2) others which are not precipitated as sulphides in acid, but only in alkaline liquids, and (3) others which are not precipitated as sulphides in either acid or alkaline solutions.

Besides this, as we have seen, some sulphides have characteristic colours, by which they are easily distinguished.

Test for Sulphuretted Hydrogen.—The smell of this gas is sufficiently characteristic to distinguish it from all others. An additional test is its action on a salt of lead, such as lead acetate. Paper moistened with a solution of lead acetate is turned black by this gas owing to the formation of lead sulphide.

EPITOME.

Sulphuretted hydrogen is prepared by the action of sulphuric or hydrochloric acid on ferrous sulphide. The gas has a disagreeable smell, like bad eggs, and is poisonous. It is moderately soluble in cold water, giving a solution having the smell of the gas. It must be collected over hot water. The gas burns with a bluish flame, giving water and sulphur dioxide if excess of air is present, but depositing some of its sulphur if the supply of air is limited.

The gas combines directly with many metals, as lead, copper, silver, forming sulphides. The brown "tarnish" which comes on silver articles exposed to the air of towns, is due to the formation of silver sulphide. Sulphuretted hydrogen also acts on metallic compounds, both in the solid state (as in the case of "white lead," oxide of iron, lime) or in solution.

On account of its behaviour towards metallic salts in solution, it is used in analysis, for separating and detecting the various metals whose compounds are present. Thus, of the six metals, *lead, copper, iron, zinc, calcium, potassium,* the sulphides of lead and copper are precipitated in *acid* solutions, whilst those of the others are not. The sulphides of iron and zinc are precipitated in *alkaline* solutions, whilst sulphides of calcium and potassium are soluble in both acids and alkalies. The action of sulphuretted hydrogen on solutions of salts of these metals is represented by the following equations—

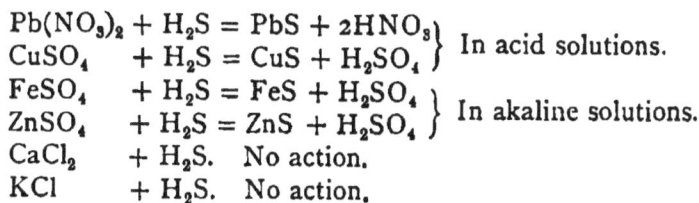

$$Pb(NO_3)_2 + H_2S = PbS + 2HNO_3$$
$$CuSO_4 \quad + H_2S = CuS + H_2SO_4 \Big\} \text{ In acid solutions.}$$
$$FeSO_4 \quad + H_2S = FeS + H_2SO_4$$
$$ZnSO_4 \quad + H_2S = ZnS + H_2SO_4 \Big\} \text{ In akaline solutions.}$$
$$CaCl_2 \quad + H_2S. \quad \text{No action.}$$
$$KCl \quad + H_2S. \quad \text{No action.}$$

CHAPTER XXVIII

SULPHUR DIOXIDE—SULPHUROUS ACID—SULPHITES.

THE two most important oxides of sulphur are sulphur dioxide, SO_2, and sulphur trioxide, SO_3. Both of these compounds are acid-forming oxides. The first, when dissolved in water, gives sulphurous acid, H_2SO_3; whilst the second gives sulphuric acid, H_2SO_4.

Sulphur Dioxide, SO_2.—This compound is always produced when sulphur is burnt in the air or in oxygen, so that we can make it this way—

$$S + O_2 = SO_2.$$

Experiment 236.—Place a piece of sulphur in a short horizontal piece of combustion tube, attach a delivery tube to one end, and arrange to pass oxygen in at the other. Heat the sulphur until it begins to burn, and let oxygen *slowly* stream through. Collect the gaseous product by downward displacement, covering the mouth of the cylinder with a piece of paper or card. [Note that the gas is not clear. This is because there is always produced at the same time a small quantity of sulphur *trioxide*, along with the *dioxide*.] Test the gas with a lighted taper, notice that the gas does not burn, and that it puts out the taper.

Fan a little of the gas towards the face, so as to get a slight whiff of it. Note its choking smell, familiarly known as "the smell of burning sulphur."

Enormous quantities of sulphur dioxide are made in this way for the manufacture of sulphuric acid. Sometimes sulphur itself, and sometimes iron pyrites, is used; and it is burnt, not in glass tubes in a stream of pure oxygen, but in special furnaces and in ordinary air.

Preparation. — When we want sulphur dioxide for experiments in the laboratory, we always make it by acting on copper with sulphuric acid.

Experiment 237.—Place a quantity of scrap copper in a flask, fitted with a cork and exit tube, and pour upon it enough strong sulphuric acid to well cover it. Heat the acid carefully with a rose burner. Notice that, as the acid gets warm, it becomes muddy, owing to the formation of a black powder, which collects in the flask in considerable quantity. Presently effervescence sets in, and sulphur dioxide is rapidly evolved. Collect four jars of the gas by displacement.

The final result of the action of sulphuric acid on copper is expressed by the equation—

$$Cu + 2H_2SO_4 = CuSO_4 + 2H_2O + SO_2.*$$

Properties of Sulphur Dioxide.—The specimens collected show that it is a colourless gas : and its smell will have been perceived during the preparation of it. We have also seen, by Exp. 236, that the gas does not burn, nor support the combustion of a taper.

Experiment 238.—Place one of the jars of gas mouth downwards in water. Notice that the gas is absorbed fairly rapidly. Add a few drops of litmus to the water and note that the solution of this gas is strongly acid. At the common temperature, water dissolves about fifty times its own volume of sulphur dioxide.

Experiment 239.—Pour into a second jar of the gas 3 or 4 cc. of litmus solution, and shake it up. As before, the litmus is instantly reddened, but after a time the colour gets fainter and finally almost entirely disappears. The gas has bleaching properties. Now add to the bleached (or nearly bleached) liquid a few drops of strong sulphuric acid. Notice that the red colour is restored.

Sulphur dioxide is used for bleaching straw, flannel, sponges, and other articles which would be injured by chlorine. Its

---

* But this does not explain everything that goes on, because it only tells us that copper sulphate, water, and sulphur dioxide are formed, and takes no notice of the black substance which is produced in the flask. Copper sulphate is not black. The equation, therefore, only gives us the *final* products. We do not know exactly what intermediate products are formed, and therefore cannot express their formation by an equation.

bleaching power is not due to the oxidation of the colouring matter, as in the case of chlorine, and is not always permanent. Sponges, flannels, and straw articles, gradually return to their unbleached state.

Flowers, especially those of a violet or purplish tint, are quickly bleached if placed in sulphur dioxide.

Although sulphur dioxide does not support the combustion of ordinary combustibles, some things will burn in this gas.

Experiment 240.—Place a small heap of lead dioxide on a deflagrating spoon, gently warm it and then lower it into a jar of sulphur dioxide. The two dioxides combine with so much energy that the lead dioxide becomes red hot, and forms *lead sulphate.*

$$PbO_2 + SO_2 = PbSO_4.$$

Experiment 241.—Sprinkle a few particles of sodium dioxide (sodium peroxide) into a jar of sulphur dioxide. The sodium dioxide takes fire and burns brilliantly, forming *sodium sulphate.*

$$Na_2O_2 + SO_2 = Na_2SO_4.$$

Sulphur dioxide is used as a disinfectant, and for this purpose is generally obtained by burning sulphur.

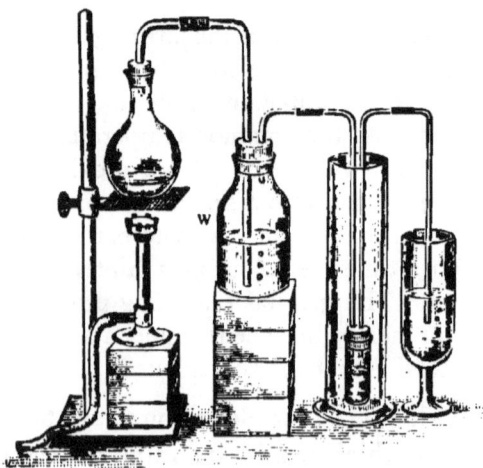

FIG. 107.

When sulphur dioxide is cooled below $-8°$ it condenses to a liquid.

Experiment 242.—Cut a test-tube about 4 cm. (nearly 2 inches) from the closed end, and border the end. Into this short tube fit a cork, carrying two long thin tubes, as shown in Fig. 107. Place this in a vessel filled with a mixture of powdered ice and salt, and connect it to the apparatus for making sulphur dioxide, first drying the gas by bubbling it through strong sulphuric acid in the bottle *w*. The ice and salt mixture cools the tubes below −8°, so that the gas which goes through will be condensed to the liquid state, and will collect in the little test-tube. In case any gas passes through without condensing, attach a delivery tube to the apparatus, and let it dip into water. This will absorb any escaping gas. When enough liquid is collected, lift the tube out of the freezing mixture, and pour the liquid upon a little water in a small dish. The sulphur dioxide boiling at −8° at once freezes some of the water. *Do not inhale the gas, as it is very irritating to the lungs.*

Liquid sulphur dioxide is used for the artificial production of ice on a large scale.

Sulphurous Acid, H_2SO_3.—A solution of this acid is produced when sulphur dioxide is dissolved in water. The solution is strongly acid towards litmus, and smells like the gas. When it is warmed the gas is driven off again, and the whole of it is expelled by boiling the solution. Sulphurous acid does not keep. It slowly absorbs oxygen from the air, and changes into sulphuric acid. This change is expressed by the equation—

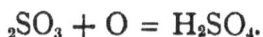

$$_2SO_3 + O = H_2SO_4.$$

On account of the readiness with which it takes up oxygen from other compounds, sulphurous acid is a powerful *reducing* substance.

Experiment 243.—Dissolve a crystal of potassium permanganate in water. This substance is very rich in oxygen ($KMnO_4$). Pour this deep violet solution into some sulphurous acid (made by passing sulphur dioxide into water). Notice that the colour is instantly discharged. The sulphurous acid takes some of the oxygen away from the permanganate, and changes into sulphuric acid.

Sulphites are salts of sulphurous acid, obtained by neutralizing the acid with a base.

Experiment 244.—Add caustic soda solution cautiously to some sulphurous acid, until a drop of the liquid on a glass rod just turns reddened litmus paper blue. Then add one or two drops more of the acid so as to make the solution *just* acid. Evaporate the solution to dryness, and obtain a white salt. This is sodium sulphite.

Add a few drops of sulphuric acid to it. Notice that there is effervescence. Smell the gas, and note that it is sulphur dioxide. Sulphuric acid, therefore, decomposes *sulphites*, expelling sulphur dioxide, and forming *sulphates*. The reactions here are—

(1) $H_2SO_3 + 2NaHO = Na_2SO_3 + 2H_2O$.

(2) $Na_2SO_3 + H_2SO_4 = Na_2SO_4 + H_2O + SO_2$.

Sulphurous acid contains two atoms of hydrogen ; it is, therefore, called a *dibasic* acid. It can form two classes of salts (just as carbonic acid does, see p. 227), depending on whether both or only one of these hydrogen atoms are displaced. Thus, there are two sodium sulphites—

(1) Normal sodium sulphite, Na_2SO_3.

(2) Hydrogen sodium sulphite (or sodium bi-sulphite), $HNaSO_3$.

Salts, like this hydrogen sodium sulphite, in which only a part of the hydrogen originally present in the acid has been exchanged, are sometimes called *acid* salts. It does not follow, however, that they have an acid reaction towards litmus, although in many cases they do happen to possess this property. It must be remembered that an *acid salt* simply means a salt in which the whole of the hydrogen of the acid has not been displaced by the metal.

Sulphites are all decomposed by stronger acids, such as hydrochloric or sulphuric.

The action of sulphuric acid is shown by the equation above. With hydrochloric acid the only difference is that a chloride of the metal is formed, thus—

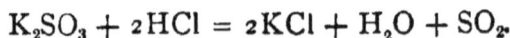

$$K_2SO_3 + 2HCl = 2KCl + H_2O + SO_2.$$

Note that when lead dioxide and sodium dioxide were burnt in sulphur dioxide, the *sulphates* and not the *sulphites* of the metals were produced.

Sulphur Trioxide, SO_3.—When sulphur burns in oxygen, however much oxygen there may be, the compound produced is always sulphur dioxide, and only a minute trace of the trioxide is formed at the same time. But if a mixture of sulphur dioxide and oxygen is passed through a heated tube containing very finely divided platinum, then the sulphur dioxide combines with the oxygen and gives sulphur trioxide. The way in which the heated platinum causes these two gases to unite together is not clearly known, but the platinum itself is not altered.

Experiment 245.—Dip some asbestos fibres into a solution of platinum chloride, and then hold them by means of a small pair of tongs in a bunsen flame until they are quite hot. The platinum chloride first dries and then decomposes, leaving the asbestos coated over with very finely divided platinum.

Now pack a quantity of this " platinized asbestos " into a bulb tube, which is supported as shown in Fig. 38. Remove the bottle *w* from the sulphur dioxide apparatus (Fig. 107), and replace it by a similar bottle having a third tube, which dips into the acid. Connect this tube with a supply of oxygen, so as to let both sulphur dioxide and oxygen bubble through the same bottle and become mixed. Notice that so long as the platinum is cold, no fumes of sulphur trioxide escape from the bulb tube, but as soon as it is heated, white fumes make their appearance. If these fumes are passed through a U-shaped tube, kept cold by being placed in a freezing mixture (powdered ice and salt), white silky-looking crystals will condense in the cold tube. This is the sulphur trioxide.

Sulphur trioxide has a most powerful affinity for water. If exposed to the air, it soon takes enough moisture from the air to convert itself into sulphuric acid. If the crystals of sulphur trioxide are dropped into water they combine with great energy, making a hissing sound like a red hot iron going into water. If placed upon the skin it produces painful burns.

Sulphuric Acid, H_2SO_4, is the most important of all the sulphur compounds, and its manufacture is carried on on an enormous scale.

The frequency with which we have used this powerful acid substance in order to bring about chemical decompositions cannot fail to have been noticed. It was used in the preparation of hydrogen, chlorine, hydrochloric acid, nitric acid, carbon monoxide, and sulphur dioxide.

Modes of Formation.—Sulphuric acid is produced when sulphur trioxide is dissolved in water.

$$SO_3 + H_2O = H_2SO_4.$$

It is also produced *slowly*, by the gradual absorption of oxygen by a solution of sulphurous acid (see Sulphurous Acid).

The process by which it is prepared on a manufacturing scale, consists in making sulphur dioxide combine with oxygen (from the air) in the presence of water (steam).

$$SO_2 + O + H_2O = H_2SO_4.$$

We have seen already that sulphur dioxide and oxygen do not combine very readily, but require help in order to make them unite. Therefore, if sulphur dioxide, oxygen, and steam were simply mixed together, they would be a very long time in uniting to form sulphuric acid.

The compound employed to cause the sulphur dioxide to combine with oxygen is nitric oxide, NO.

We have learnt already (p. 199) that when nitric oxide comes in contact with the air, it unites with another atom of oxygen, and forms nitrogen peroxide, NO_2, a reddish gas. Now nitrogen peroxide easily gives up this extra atom of oxygen to sulphur dioxide in the presence of steam, and goes back again to nitric oxide. Therefore, when steam, sulphur dioxide, and nitrogen peroxide are mixed, the following change takes place —

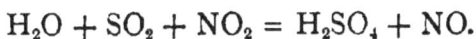

$$H_2O + SO_2 + NO_2 = H_2SO_4 + NO.$$

The nitric oxide that is thus formed instantly seizes another atom of oxygen from the air, again forming nitrogen peroxide, NO_2,

$$NO + O = NO_2$$

and this again hands on this extra oxygen to another portion of sulphur dioxide.

The nitric oxide (NO) is, therefore, a sort of "middleman," who takes oxygen from the air and passes it on to the sulphur dioxide ; and the same quantity of nitric oxide can keep on doing this, and can convert an unlimited amount of sulphur dioxide into sulphuric acid.

The Manufacture of Sulphuric Acid is carried on in enormous rooms made of sheet lead. These great rooms are called *leaden chambers*, and they are often of such a size that 250 people could sit down to dine in one. Generally several are placed in a row.

Into these chambers are sent sulphur dioxide, air, nitrogen peroxide, and steam.

(1) The sulphur dioxide is produced either by burning sulphur, or roasting iron pyrites, in special furnaces called *sulphur burners* or *pyrites burners*. In either case sulphur dioxide is formed, which, with the excess of air passing through the furnace, is drawn into the "chambers."

$$S + O_2 = SO_2.$$

Sometimes the sulphur dioxide is obtained by burning sulphuretted hydrogen.

(2) The nitrogen peroxide is produced by placing inside one of the sulphur burners a pot containing a little Chili saltpetre and sulphuric acid. This mixture when heated gives nitric acid (p. 189), and the fumes of nitric acid coming in contact with the sulphur dioxide are decomposed, yielding nitrogen peroxide, which passes on into the chambers.

$$2\,HNO_3 + SO_2 = H_2SO_4 + 2\,NO_2.$$

S

(3) The air that is admitted into the chambers is what is allowed to pass through the pyrites burners, and its amount is regulated.

(4) The steam is blown into the chambers in jets from steam boilers.

When these gases mix in the chambers, the chief reactions which go on are represented by the equations given above, and sulphuric acid (moderately strong) collects on the floors and is drawn off. Care is taken to adjust the proportion of the various gases.

As the oxides of nitrogen go on transferring oxygen from the air to the sulphur dioxide over and over again, it is only necessary to add a very small amount of the nitrogen peroxide from time to time, to make up for the slight loss of this gas which always takes place.[1]

The acid which is drawn from the chambers (*chamber acid*) is boiled down, either in glass or platinum vessels, so as to drive off the water, and so get strong sulphuric acid.

Properties.—Pure sulphuric acid is a heavy, colourless, oily liquid (hence the common name *oil of vitriol*). It is a powerfully corrosive substance, and if spilt upon the skin produces bad burns. Therefore some care must be taken in handling this acid.

It has a strong affinity for water, and if mixed with water the mixture gets nearly boiling hot. On account of its power of combining with water, it is constantly used for withdrawing water vapour from gases. Thus, when we require to dry a gas, that is, to remove the vapour of water from it, we bubble the gas through sulphuric acid, provided it is a gas which has no action on the acid. If this acid is exposed to the air, or is not kept in well-stoppered bottles, it quickly absorbs water vapour from the air, and, of course, by so doing gets more and more dilute.

Its affinity for water is so great that it decomposes many compounds containing hydrogen and oxygen, and takes these elements away from the compound in the proportion to yield

[1] For further details of the manufacturing process, see "Newth's Inorganic Chemistry."

water.   Thus, in the case of oxalic acid (p. 233), $C_2H_2O_4$. This is decomposed by sulphuric acid, which in its eagerness for water, abstracts from the oxalic acid the elements which yield water, H_2 and O, leaving just enough oxygen for the carbon to form carbon monoxide and carbon dioxide.

Again in the case of sugar (p. 212), $C_{12}H_{22}O_{11}$. When sulphuric acid acts on this, it abstracts all the hydrogen and oxygen (which are present in exactly the proportion to give $11H_2O$) and leaves the carbon in the free or uncombined state.   Its power of charring organic matter may be shown by the following experiment.

Experiment 246.—Take some very dilute sulphuric acid, and with the finger write a word on a piece of paper.   Now gently dry the paper by holding it at some distance above a gas flame.   Notice that the paper is charred where the letters were drawn upon it.

Sulphates.—Just as carbonic acid and sulphurous acid form two classes of salts, so, for the same reason, there are two classes of sulphates.   The reason being that sulphuric acid, like these others, contains two atoms of hydrogen which can be replaced by metals.

Thus, by replacing the hydrogen atoms by potassium we get either normal potassium sulphate (or potassium sulphate), K_2SO_4; or hydrogen potassium sulphate (potassium bi-sulphate), $HKSO_4$.

Certain of the sulphates were known to the very early chemists, and were called *vitriols* (because they had rather a vitreous or glassy appearance), such as blue vitriol (*copper sulphate*), green vitriol (*iron sulphate*), white vitriol (*zinc sulphate*).   The name " oil of vitriol " is derived from the fact that the acid was formerly obtained by distilling green vitriol.

The sulphates, like all other salts, are formed when the acid is neutralized with a metallic hydroxide.

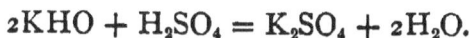

$$2KHO + H_2SO_4 = K_2SO_4 + 2H_2O.$$

They are also produced by the action of the acid on oxides and carbonates—

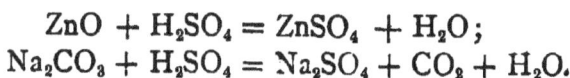

$$ZnO + H_2SO_4 = ZnSO_4 + H_2O;$$
$$Na_2CO_3 + H_2SO_4 = Na_2SO_4 + CO_2 + H_2O.$$

Being such a powerful acid, sulphuric acid is capable of taking metals away from the salts of almost any other acid; for example, from sodium chloride or potassium nitrate it takes the sodium or potassium, and gives its hydrogen in exchange.   Thus —

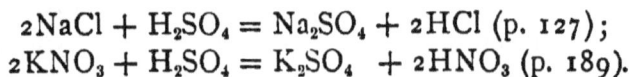

$$2NaCl + H_2SO_4 = Na_2SO_4 + 2HCl \text{ (p. 127)};$$
$$2KNO_3 + H_2SO_4 = K_2SO_4 + 2HNO_3 \text{ (p. 189)}.$$

Test for Sulphates.—(1) Sulphates which are soluble in water are recognized and distinguished by giving a white precipitate with barium chloride, consisting of barium sulphate. This white precipitate is *insoluble in either hydrochloric or nitric acid.*

Experiment 247.—Dissolve in three separate test-tubes a small particle of (1) potassium sulphate, (2) sodium sulphite, and (3) sodium carbonate.   Add to each a few drops of barium chloride. Notice that a very similar precipitate is formed in each case, but in reality they are totally different—one is barium *sulphate*, the next is barium *sulphite*, and the third is barium *carbonate*.   To each add a few drops of strong hydrochloric acid.   Notice that the barium sulphate is unaffected ; the barium carbonate quickly dissolves with effervescence, giving off carbon dioxide ; while the barium sulphite also dissolves with effervescence, and gives off sulphur dioxide (which can be detected by the smell).   [Probably in this case the precipitate will not *wholly* dissolve, because the sodium sulphite originally used is likely to contain a little sodium *sulphate* mixed with it, so that the precipitate obtained when barium chloride was added, consists partly of barium sulphite (which will dissolve in the acid) and barium sulphate which will not dissolve.] The reactions in this experiment are the following.

(a) When barium chloride is added to the three solutions—

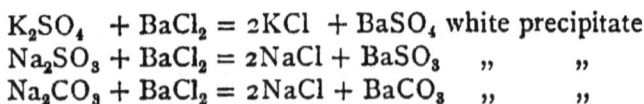

$$K_2SO_4 \quad + BaCl_2 = 2KCl \quad + BaSO_4 \text{ white precipitate}$$
$$Na_2SO_3 + BaCl_2 = 2NaCl + BaSO_3 \quad ,, \qquad ,,$$
$$Na_2CO_3 + BaCl_2 = 2NaCl + BaCO_3 \quad ,, \qquad ,,$$

(b) The action of hydrochloric acid on the three precipitates—

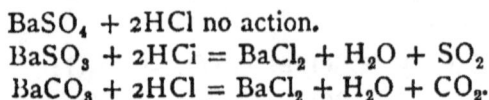

$$BaSO_4 + 2HCl \text{ no action.}$$
$$BaSO_3 + 2HCi = BaCl_2 + H_2O + SO_2$$
$$BaCO_3 + 2HCl = BaCl_2 + H_2O + CO_2.$$

(2) Sulphates which are not soluble in water are tested for in a different way.

Experiment 248.—Take a pinch of plaster of Paris (calcium sulphate) and mix it with about three times as much powdered sodium carbonate, and heat the mixture on a little piece of platinum foil, bent into a sort of spoon (or on the lid of a platinum crucible) until it has completely melted. At this high temperature the two compounds make a mutual exchange, resulting in the formation of sodium sulphate and calcium carbonate.

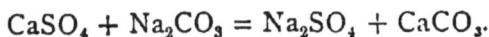

$$CaSO_4 + Na_2CO_3 = Na_2SO_4 + CaCO_3.$$

The sodium sulphate is soluble in water, while the calcium carbonate is not ; therefore, put the platinum spoon into water in a test-tube and boil it for a minute or two, and then filter it. The solution contains sodium sulphate, *and any excess of sodium carbonate which may have been used.* Now add hydrochloric acid until the solution is quite acid, and then add barium chloride. The white precipitate of barium sulphate is at once formed. [Instead of *fusing* the insoluble sulphate with sodium carbonate, we may mix it with a strong solution of sodium carbonate and boil it for some time, when there will be enough of the calcium sulphate decomposed and sodium sulphate produced to give the test with barium chloride.]

Fig. 108.

Experiment 249.—Heat a little heap of a mixture of plaster of Paris and sodium carbonate on a piece of charcoal by means of a small blow-pipe flame, as shown in Fig. 108. [Select a good piece of charcoal, which is not full of cracks, and scoop a small hollow upon it to hold the substance which is being heated.] The result of this operation, is that the sodium of the sodium carbonate combines with the sulphur of the calcium sulphate to give *sodium sulphide.* When it is cold, place the fused residue on a silver coin and touch it with a drop of water. The sodium sulphide at once acts on the silver and causes a black stain upon it.

EPITOME.

The two commonest oxides of sulphur are sulphur dioxide, SO_2, and sulphur trioxide, SO_3. Sulphur dioxide is a colourless choking gas, obtained when sulphur burns either in air or in oxygen. It is prepared by heating sulphuric acid and copper. It is more than twice as heavy as air, and therefore can be collected by downward displacement.

It is soluble in water, and therefore cannot be collected at the pneumatic trough.

Sulphur dioxide does not burn, nor support the combustion of ordinary burning bodies. Lead dioxide and sodium dioxide take fire in the gas and produce sulphates of the metals.

Sulphur dioxide bleaches, but not always permanently; the colour often being restored either by stronger acids or by alkalies.

The gas is used for disinfecting purposes; but its chief use is in the manufacture of sulphuric acid.

Sulphur dioxide is easily condensed to a liquid by cooling it. The liquefied gas is colourless, and boils at $-8°$.

The solution of sulphur dioxide in water is acid, and contains sulphurous acid, H_2SO_3. This acid has never been obtained except as a solution in water. When the solution is boiled, the acid is decomposed and sulphur dioxide escapes.

The salts of sulphurous acid are called sulphites. The acid is dibasic, and therefore forms two classes of salts; those in which all the hydrogen has been replaced by metals, and those in which only half the hydrogen is so replaced. Thus : Na_2SO_3, di-sodium sulphite, or normal sodium sulphite ; and $HNaSO_3$, hydrogen sodium sulphite (*sometimes called acid sodium sulphite, or sodium bisulphite*).

The sulphites are decomposed by hydrochloric or sulphuric acid, with the evolution of sulphur dioxide.

Sulphur trioxide is a white solid, forming silky crystals. It is produced when sulphur dioxide and oxygen are passed over heated spongy platinum. This oxide has a powerful affinity for water, with which it combines to form sulphuric acid.

Sulphuric acid (or oil of vitriol) is a colourless oily liquid, strongly corrosive, and a powerful acid. Its manufacture is the most important of all chemical industries. It is made by oxidizing sulphur dioxide by means of nitrogen peroxide in the presence of steam. Sulphur dioxide combines with atmospheric oxygen, only with extreme slowness ; but, by means of nitric oxide, oxygen is taken from the air and handed on to the sulphur dioxide. Nitric oxide unites with oxygen of the air, forming nitrogen peroxide, and this gives oxygen to the sulphur dioxide, and is again reduced to nitric oxide.

The operation is carried on in enormous chambers built of lead.

Sulphuric acid combines with water with the production of great heat. On account of its eagerness to unite with water it is used for drying gases. It also decomposes many organic compounds containing oxygen and hydrogen, withdrawing these two elements in the proportion required to form water.

Sulphuric acid is dibasic, and forms two classes of salts, according as to whether all, or only half, the hydrogen of the acid is replaced by metals.

CHAPTER XXX.

SOME COMMON CARBON COMPOUNDS.

CARBON forms such an enormous number of compounds that a simple list of their names alone would more than fill this little book. The study of these compounds is generally called "organic chemistry," because, in the early days of chemistry, it was thought that these compounds could only be produced as the result of living organized bodies, like animals and plants.

In order to study this vast host of compounds, they are divided and subdivided into classes and families much in the same way as animals or plants are classified.

Four very important classes are :—

1. Hydrocarbons.
2. Acids.
3. Alcohols.
4. Carbohydrates.

1. Hydrocarbons.—These, as the name implies, are compounds of carbon with hydrogen only. Important amongst these are the following :—

Marsh Gas, CH_4.—Found in coal mines, and called *fire-damp*. Also in marshy places where vegetable matter is rotting. In the laboratory it is made by strongly heating a mixture of sodium acetate and caustic soda, when sodium carbonate is left behind.

$$NaC_2H_3O_2 + NaHO = Na_2CO_3 + CH_4.$$

Marsh gas is colourless, and has no smell. It burns easily, but gives very little light, although rather more than hydrogen.

If mixed with air, or with pure oxygen, and fired, the mixture explodes. This is the cause of coal-mine explosions. Water and carbon dioxide are produced, and the latter gas is called *choke-damp* by the miners.

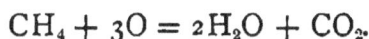

$$CH_4 + 3O = 2H_2O + CO_2.$$

Ethylene, C_2H_4 (olefiant gas), is obtained from common alcohol (*spirits of wine*) by heating it with sulphuric acid (see Exp. 184).

$$C_2H_6O - H_2O = C_2H_4$$

Ethylene is a colourless gas with a faint pleasant smell. It burns easily, with a very bright flame, producing carbon dioxide and water, the same products as are formed when any hydrocarbon burns.

Acetylene, C_2H_2, is formed whenever coal gas burns without a sufficient supply of air. Thus, when a Bunsen lamp gets alight down at the little jet at the base of the chimney, some acetylene is produced. This it is which causes the bad smell we notice when the lamp so burns.

Acetylene is best prepared in the laboratory by acting on calcium carbide with water. A little of the carbide, in small lumps, is put into a test-tube, and a few drops of water added. The gas is at once evolved, and can be lighted at the mouth of the tube.

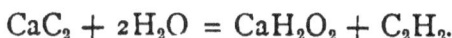

$$CaC_2 + 2H_2O = CaH_2O_2 + C_2H_2.$$

Acetylene burns with a very bright and smoky flame.

Marsh gas, Ethylene, and Acetylene are all present in ordinary coal gas.

Other important hydrocarbons are the various *mineral oils* (paraffin oils) used for illuminating purposes. *Paraffin wax*, used for candles. *Turpentine.*

2. The Acids.—This is a large and important class, divided into many families. Amongst the most important of these is the one known as the *acetic, or fatty series of acids.* These include acids present in many fats.

Formic Acid (CH_2O_2) is found in ants, and in the hairs of stinging-nettles. When we are stung by ants or nettles, it is because a minute drop of this formic acid has been injected into the skin. If a piece of litmus paper is placed upon an ants' nest which has been just disturbed, it is instantly reddened by a shower of formic acid being squirted against it by the irritated ants. Formic acid is decomposed by strong sulphuric acid into water and carbon monoxide.

Acetic Acid ($C_2H_4O_2$) is the acid of vinegar. When beer goes sour it is owing to the formation of acetic acid, and this change is brought about by the agency of a living organism familiarly known as *mother of vinegar*. Acetic acid is also produced when wood is destructively distilled, the material so obtained being called *pyroligneous acid*, that is, the *fire-wood acid*.

Pure acetic acid is liquid at the temperature of a warm room; but in winter it freezes to an ice-like solid. It is on this account called " glacial " acetic acid. This strong acid stings the skin as formic acid does. The salts are called acetates; one of the commonest is lead acetate, familiarly known as *sugar of lead*.

Butyric Acid is the name of the acid which is present in rancid butter, and which gives to it the disagreeable smell.

Palmitic Acid and **Stearic Acid** are important constituents of most solid animal fats, such as beef and mutton suet, butter, and human fat. Palmitic acid is also one of the chief constituents of palm oil (hence its name). These two acids are extensively used for making candles.

3. Alcohols. This class contains a number of very important compounds. The following may be taken as examples.

Methyl Alcohol (*wood spirit*), CH_4O or CH_3HO, is present in the watery liquid obtained when wood is distilled. When it is extracted from this liquid and purified, it is a colourless liquid, which burns with a flame without light, and without any smoke.

Ethyl Alcohol, C_2H_6O or C_2H_5HO, is the familiar *spirits of wine*. It is the best known of all the alcohols, and is therefore called simply " alcohol."

It is obtained by the fermentation of ordinary sugar or of grape sugar (glucose) by means of yeast. The yeast organism transforms the sugar into alcohol and carbon dioxide (see Exp. 195).

$$C_6H_{12}O_6 = 2C_2H_6O + 2CO_2.$$

Glucose.   Alcohol.

The liquid obtained contains only a small proportion of alcohol mixed with a large quantity of water. It has, therefore, to be distilled, or "rectified," in order to separate the alcohol from the water.

Ethyl alcohol is present in all fermented liquors, and it is the presence of this compound which gives them their intoxicating properties. Ordinary beer contains from 3 to 6 per cent. of alcohol. [If some beer is boiled in a flask with a long upright tube fastened into the neck with a cork, the alcohol which first boils off can actually be lighted as it escapes from the tube.] Light wines, such as claret, etc., contain from 8 to 14 per cent.; port and sherry 15 to 25 per cent., while brandy and other "spirits" contain from 50 to 60 per cent. of alcohol.

Alcohol is capable of dissolving things like resins, gums, oils, etc., and is therefore a most useful substance for the manufacture of varnishes, and for other purposes.

Methylated Spirit.—On account of the high duty upon "alcohol," it is too expensive for most of the manufacturing and chemical purposes for which alcohol is required. Therefore a mixture consisting of 90 per cent. "spirits of wine" and 10 per cent. "wood spirit" (impure methyl alcohol) is used instead of pure "alcohol." This mixture called *methylated spirit* is quite unfit for drinking, and, being duty free, is quite cheap.

Phenol or **Phenyl Alcohol, C_6H_5HO.**—This substance is familiarly known by the name *carbolic acid*, although in reality it is not an acid, but belongs to the class of alcohols. It is formed when coal is distilled for making coal-gas, and is extracted from the coal-tar oil.

Pure phenol forms long needle-shaped crystals, which melt a little above the temperature of the hand ($42°$). It has a sharp burning taste and is poisonous. It is a powerful disinfectant and antiseptic, and is largely used in surgery.

The common "carbolic acid powders" sold in tins for disinfecting purposes, consist of some powder such as lime or gypsum, impregnated with about 10 per cent. of very crude carbolic acid.

Glycerin, $C_3H_5(HO)_3$.—This familiar substance is also an alcohol. Chemists call it *glycerol*. It is an important constituent of fats. Fats are compounds of glycerin with such acids as palmitic and stearic. Mutton suet, for instance, consists chiefly of a compound of stearic acid and glycerin. This compound is called *stearin*. *Palmitin* is the name of the compound of palmitic acid with glycerin. Glycerin is obtained from fats by heating them in very hot steam, or by boiling them with a caustic alkali such as sodium hydroxide. The alkali combines with the stearic or palmitic acid forming a salt (soap), which separates out when the liquid cools ; while the glycerin which is set free remains dissolved in the watery liquid.

Glycerin is a thick syrupy liquid, with a very sweet taste, almost like sugar syrup.

Soap.—When fats are boiled with caustic alkalies, the fat (which is a compound of fatty acids with glycerin) is decomposed ; glycerin is set free and the fatty acid combines with the alkali. This process is called *saponification ;* and the compound of the fatty acids (chiefly stearic and palmitic acids) with the alkali, is known as a *soap.*

When caustic potash is the alkali used *soft soap* is produced ; while if caustic soda is employed the soap is harder, as in the ordinary forms of soap used for washing purposes.

In some out of the way parts of the world, people often make a crude kind of soap by collecting the ashes of burnt wood (these contain potash, the name potash simply meaning pot-ashes), mixing them with water, and boiling the liquid so obtained with mutton or beef fat

The Action of Hard Waters upon Soap.—The property of *hardness* in water is chiefly due to the presence of either carbonate of lime or sulphate of lime (see p. 81).

When water containing these salts in solution comes in contact with a solution of soap (say sodium stearate), a chemical action takes place, thus —

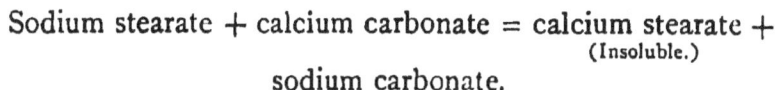

Sodium stearate + calcium carbonate = calcium stearate +
(Insoluble.)
sodium carbonate.

The calcium stearate, being insoluble, separates out as a greasy scum, which is always seen when hard water is used for washing. So long as any of the lime salts remain in the water, the soap is used up in bringing about this double decomposition, and, therefore, is wasted so far as its powers of cleansing are concerned; for we cannot wash with soap until the water has become softened by the removal or destruction of the hardening lime salts. As soon as ever all the calcium carbonate or sulphate has been decomposed by the soap, then, and not till then, will the soap give a *lather;* and not until a lather can be produced is the soap of any service for washing.

Experiment 250.—Take two good-sized stoppered bottles, and about half fill one with some hard water (common tap-water will generally do, but if this should happen to be a soft water, a sample of hard water can readily be made by adding a little lime-water, and then bubbling carbon dioxide through it). At first, calcium carbonate $CaCO_3$ is precipitated, but presently this dissolves in the carbonic acid, forming the bi-carbonate of calcium $H_2Ca(CO_3)_2$, or $CaCO_3,H_2CO_3$. Into the other bottle put an equal quantity of distilled water, or rain water. Now make a solution of soap by shaking up a few thin shavings of soap with a little water in a bottle. Add a small quantity of this soap solution at a time to the soft water, until on shaking the bottle a lather is raised which does not disappear again. Now add the soap solution to the hard water. Note that at first no lather is produced, although much more of the soap is added. Also observe that the solution turns muddy, and a scum is produced. This is the calcium stearate being precipitated; continue adding the soap until at last a permanent lather is obtained. If a little more hard water or a few drops of a solution of any lime salt be now added, the lather instantly disappears.

4. Carbohydrates.—This family of carbon compounds includes the sugars, starches, etc. The following are some common examples of carbohydrates.

Cane-sugar (*Saccharose*), $C_{12}H_{22}O_{11}$. This is ordinary sugar, and is obtained chiefly from sugar-cane and from beet-root. It is present in almost all sweet fruits.

When cane-sugar is brought into contact with yeast, it is converted into dextrose and lævulose. The same change takes place when it is warmed with dilute sulphuric acid.

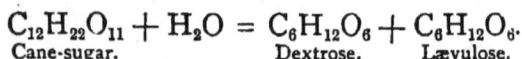

$$C_{12}H_{22}O_{11} + H_2O = C_6H_{12}O_6 + C_6H_{12}O_6.$$
Cane-sugar.   Dextrose.   Lævulose.

Grape-sugar (*Glucose* or *Dextrose*), $C_6H_{12}O_6$, is obtained from the juice of sweet grapes. It is also present in honey. It is not so sweet as cane-sugar, and is less easily dissolved by water. Grape-sugar is distinguished from cane-sugar by the following test :—

Experiment 251.—Take a crystal of copper sulphate, and one about the same size of tartaric acid. Dissolve them together in a little water, and add a solution of caustic potash until the liquid is strongly alkaline. Add a little of this mixture to a solution of grape-sugar in a test-tube, and warm the liquid. A red precipitate of cuprous oxide is formed, and the solution loses its blue colour. Do the same with a solution of common sugar ; this does not give the red precipitate, and the liquid remains blue.

Starch, $C_6H_{10}O_5$, occurs in many parts of plants, such as the seeds, stems, roots, and tubers, and it may be obtained from them by crushing them with water, and separating the bruised fibre or pulp by means of a sieve.

Experiment 252.—Take a raw potato and rub it down on a coarse grater, and collect all the pulp on a piece of muslin. Screw it up into a sort of bag, and squeeze it into a small basin of water ; dipping it once or twice into the water and squeezing again. In this way the starch passes through the muslin into the water, which is thereby made milky. If this is allowed to stand, the clean white starch settles to the bottom. If wheat meal or flour is treated in a similar manner, starch is also separated, and a whitish sticky mass is left. This is called *gluten*. It is a compound containing nitrogen, and it is the presence of this substance which gives to the wheat its value as a food. Wheat flour contains about $\frac{1}{10}$th of its weight of gluten.

When starch is boiled with dilute sulphuric acid, or is acted on by the ferment present in germinating barley (*diastase*), it is converted into glucose and dextrin.

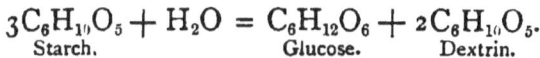

$$3C_6H_{10}O_5 + H_2O = C_6H_{12}O_6 + 2C_6H_{10}O_5.$$
Starch.   Glucose.   Dextrin.

When examined through a microscope, starch is found to consist of minute granules, not crystals. The size and shape of these vary considerably. Those of potato starch and arrowroot are much larger than those from rice.

Starch gives a blue colour with iodine (see p. 134).

Starch does not dissolve in cold water, but with boiling water the granules swell up and burst. If the amount of hot water is very large, the starch disappears and seems to dissolve, but with less water it forms a gelatinous or pasty mass.

Dextrin.—When starch paste is boiled with a little dilute sulphuric acid, it soon becomes thinner and thinner, being changed into glucose and dextrin, both of which are soluble. Dextrin is a sticky gummy substance, often used instead of ordinary gum or paste, for mounting photographs and other similar purposes. It is sold under the name of *British gum.*

CHAPTER XXXI.

SIMPLE QUALITATIVE ANALYSIS.

THE word "analysis" means the breaking up or separation of a compound into its components or elements (see p. 75). But the word is also used in a broader sense, and is applied to any processes or methods by which the chemist is able to find out what a substance is composed of, in order to identify that substance. For example, we speak of "*microscopic analysis*," and *spectrum analysis;* these are not processes in which compounds are split up into their components, but are methods which enable chemists to identify different substances. The following illustration will make this plain.

Experiment 253.—Take a crystal of nitre (potassium nitrate), and just touch the edge of a Bunsen flame with it. Notice the lilac or violet colour which it imparts to the flame. This colour is characteristic of all potassium compounds (compare Exp. 44). Place the crystal on a glass plate, dissolve it in a drop of warm water, and allow the solution to evaporate. Examine the crystals which are formed with a pocket lens or microscope. Notice the long prismatic shaped crystals, characteristic of nitre. Dissolve a crystal of the salt in water, and apply the test for a nitrate, as explained on p. 195.

By these three experiments or tests, we have *analysed* this substance, and identified it as potassium nitrate, although we have not *separated* the compound into its constituent elements.

Separation Usually Necessary.—In a great many instances, the tests which are used to identify a substance are interfered with if certain other substances are present at the same time. In all such cases it is necessary to *separate* the substances before applying the special test. For example—

Experiment 254.—Dissolve a small crystal of potassium chloride in a drop or two of water, and add to it one or two drops of a solution of platinum chloride ($PtCl_4$). A precipitate is produced, consisting of tiny yellow crystals. This is a characteristic test for potassium compounds.

Now treat a similar quantity of ammonium chloride in exactly the same way; notice that a precisely similar looking yellow crystalline precipitate is formed.

Therefore, before applying this test for the detection of a potassium salt, it is absolutely necessary to separate it from any ammonium salts.

How Separation is made.—The method most frequently used for bringing about analytical separation, is to cause one or more of the compounds present to undergo a double decomposition with certain chosen reagents, whereby the metals in these compounds form fresh compounds *which are insoluble* in water, and which are, therefore, thrown down as precipitates. An example of this method is given on p. 20, Exp. 25. The silver nitrate present is caused to enter into double decomposition with sodium chloride. This results in the formation of silver chloride, which is precipitated as an insoluble compound, and sodium nitrate remains in solution along with the copper salt. It must be noted that in this separation it is only the *silver* from the silver nitrate that is actually separated from the copper compound, for the other part of the compound is left in solution combined with *sodium* instead of with silver. By this process we have not separated *silver nitrate* from copper nitrate, but only withdrawn the silver and replaced it by sodium.

Reagents.—The solutions which are used to separate substances in this way are called *reagents*. Sometimes the same reagent will form insoluble compounds with a whole group of metals, in which case it may be used to separate an entire family of metals from others not belonging to the group. Such reagents are called *group-reagents*.

T

Groups.—For convenience the metals are divided into a number of groups, based upon their behaviour towards certain chosen group-reagents, used in a certain order.

GROUP I. or HYDROCHLORIC ACID GROUP.—Metals whose chlorides are precipitated on the addition of hydrochloric acid.

Lead [1] (silver, mercury).

GROUP II. or SULPHURETTED HYDROGEN GROUP.— Metals whose sulphides are precipitated from *acid* solutions by sulphuretted hydrogen.

Lead, [2] **Copper** (mercury, bismuth, cadmium, tin, arsenic, antimony).

GROUP III*a.* or AMMONIA GROUP.—Metals whose hydroxides are precipitated by ammonia, in the presence of ammonium chloride.

Iron (chromium, aluminium).

GROUP III*b.* or AMMONIUM SULPHIDE GROUP.—Metals whose sulphides are precipitated by ammonium sulphide, in the presence of ammonia.

Zinc (manganese, nickel, cobalt).

GROUP IV. or AMMONIUM CARBONATE GROUP.—Metals whose carbonates are precipitated by ammonium carbonate, in the presence of ammonium chloride.

Calcium (barium, strontium).

GROUP V.—No group reagent.

Potassium, Ammonium (sodium, magnesium).

Reactions for the Metals.—In order that we may be able to recognize and identify a metal in the various compounds it produces with different reagents, it is necessary to make ourselves quite familiar with these compounds. In order to gain this knowledge, the following reactions or tests should be carefully made, and the student should make exact notes of all he does and observes. If any experiment he makes seems to give a different result from that which is

---

[1] Those metals printed in thick type are the only ones the elementary student will be concerned with.

[2] The reason why lead is in both groups I. and II. is explained on page 276.

indicated in the book, he should not pass it over, but should repeat it more carefully.

Lead, Pb. (Group I.)

Use lead nitrate, $Pb(NO_3)_2$. Take a few crystals of the salt; note their dry, hard, milk-white appearance. Dissolve a little in water in a test-tube. Note that it is not very readily dissolved in cold water, but more quickly if the water is warmed. Pour some of this solution into five separate test-tubes.

I. To one, add a few drops of dilute **hydrochloric acid** (group-reagent). Note the white, curdy precipitate of lead chloride. Shake the tube, and, when the precipitate has settled, add more of the reagent, until no further precipitation takes place.

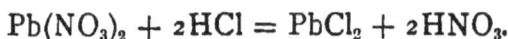

$$Pb(NO_3)_2 + 2HCl = PbCl_2 + 2HNO_3.$$

Now heat the mixture, and observe how the precipitate · disappears. Lead chloride is soluble in hot water. *This distinguishes lead from the other metals of this group.* Cool the test-tube again, and the lead chloride is again precipitated ; but note that it comes down in the form of shining white needle-shaped crystals.

II. Take the second portion of the lead nitrate solution, and pass sulphuretted hydrogen through it (see p. 246), or add sulphuretted hydrogen water. A black precipitate of lead sulphide is formed.

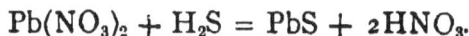

$$Pb(NO_3)_2 + H_2S = PbS + 2HNO_3.$$

Filter the liquid, scrape a little of the precipitate off the filter into a test-tube, add a few drops of strong nitric acid, and boil. Notice that the black precipitate turns white. The lead sulphide is oxidized by the nitric acid into lead sulphate.

III. Treat the third portion as the first, and filter the mixture. Then pass sulphuretted hydrogen into the clear filtrate. Note a black precipitate, as in II. This shows that the group reagent does not *entirely* separate lead from group II.; that is to say, lead chloride is slightly soluble even in cold water, and, therefore, a portion of it passes through along with the metals of group II. (see p. 274).

IV. To the fourth portion add potassium chromate. Note the yellow precipitate of lead chromate.

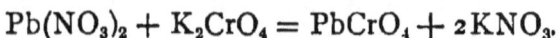

$$Pb(NO_3)_2 + K_2CrO_4 = PbCrO_4 + 2KNO_3.$$

V. To the fifth portion add dilute sulphuric acid; a white granular precipitate of lead sulphate is produced.

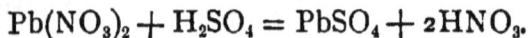

$$Pb(NO_3)_2 + H_2SO_4 = PbSO_4 + 2HNO_3.$$

Dry Reaction.—Powder a small crystal of lead nitrate, mix about twice as much sodium carbonate with it, and place the mixture in a shallow cavity scooped out on a piece of charcoal. Heat the mixture by means of a blow-pipe flame (see Fig. 108, p. 261), holding the charcoal in such a position that the *middle or inner* part of the flame, and not the tip of it, plays upon the mixture. [The outer part of the flame, where the oxygen of the air is in excess, is called the *oxidizing flame;* while the inner portion, where heated coal-gas is in excess, is known as the *reducing flame.*]

The mixture quickly melts, and the lead compound is reduced to the state of metallic lead, which will appear in the form of small brilliant globules upon the charcoal. At the same time, some of the lead is oxidized, and oxide deposits round the cavity as a yellow incrustation. Allow the mass to cool; pick out one of the globules of metal with a penknife, and show that it is soft and malleable by hammering it; also that if rubbed across paper it leaves a black mark.

Copper, Cu. (Group II.)

Use copper sulphate, $CuSO_4,5H_2O$. Dissolve a little of the salt in water. Note that it is readily soluble.

I. Take a small portion of the solution and pass **sulphuretted hydrogen** (the group-reagent). Notice the brownish-black precipitate of copper sulphide.

$$CuSO_4 + H_2S = CuS + H_2SO_4.$$

Filter the mixture, and show that the precipitate dissolves in boiling nitric acid, giving a bluish solution. (Compare the behaviour of lead sulphide.) Cautiously add ammonia to this solution, and note the deep azure-blue colour of the liquid.

II. To a second portion of the copper sulphate solution add ammonia, drop by drop. Notice the pale greenish-blue precipitate. Add more ammonia, and the precipitate quickly dissolves, forming a deep blue solution, characteristic of copper compounds.

Iron, Fe. (Group IIIa.)

This element forms two classes of compounds, which behave quite differently towards reagents. One of these classes is derived from *ferrous oxide*, FeO, and the other from *ferric oxide*, Fe_2O_3, in which the iron is combined with a larger proportion of oxygen, or is in a *higher state of oxidation*, as we say. These two classes of iron compounds are therefore distinguished as *ferrous* and *ferric* salts. The former are mostly pale green (Exp. 51, p. 46) or white, while the ferric salts are generally yellow.

Ferrous are readily converted into *ferric* salts by the action of oxidizing agents ; while reducing agents change *ferric* back to *ferrous* compounds. Thus, if sulphuretted hydrogen is passed through a neutral or acid solution of *ferric* chloride, *ferrous* chloride is produced and sulphur is precipitated.

$$Fe_2Cl_6 + H_2S = 2FeCl_2 + 2HCl + S.$$

Use ferrous sulphate, $FeSO_4,7H_2O$, and ferric chloride, Fe_2Cl_6.

Make a solution of ferrous sulphate by dissolving two or three crystals of the salt in *cold* water.

I. Add **ammonia** (group-reagent), and obtain a dirty greenish precipitate of ferrous hydrate, which on exposure to the air, gradually becomes oxidized into *ferric* hydrate (brown).

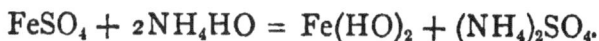

$$FeSO_4 + 2NH_4HO = Fe(HO)_2 + (NH_4)_2SO_4.$$

Repeat with a solution of ferric chloride ; a reddish-brown precipitate of ferric hydrate is formed.

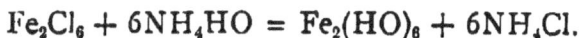

$$Fe_2Cl_6 + 6NH_4HO = Fe_2(HO)_6 + 6NH_4Cl.$$

II. Add ammonium sulphide to another portion of the ferrous sulphate solution. Black ferrous sulphide is precipitated.

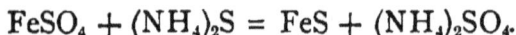

$$FeSO_4 + (NH_4)_2S = FeS + (NH_4)_2SO_4.$$

Repeat with ferric chloride. The same precipitate is obtained, mixed with sulphur.

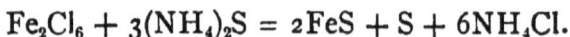

$$Fe_2Cl_6 + 3(NH_4)_2S = 2FeS + S + 6NH_4Cl.$$

III. Add a few drops of ammonium thiocyanate (frequently wrongly called ammonium sulphocyanide) to the ferrous sulphate. No change takes place.

Repeat with ferric chloride. An intense blood-red coloured solution is produced.

IV. Add a few drops of potassium ferricyanide to the ferrous sulphate. A dark blue precipitate is formed (called Turnbull's blue).

Repeat with ferric chloride. No blue colour, but the mixture becomes brownish.

V. Add potassium ferrocyanide to ferrous sulphate, a light blue precipitate is formed, which on exposure to air becomes darker blue.

Repeat with ferric chloride. A dark blue precipitate results (called Prussian blue).

Zinc, Zn. (GROUP III*b*.)

Use zinc sulphate, $ZnSO_4, 7H_2O$. Dissolve a few crystals in water. Note the appearance of the crystals and their ready solubility.

I. Add **ammonium sulphide** (group-reagent). A white precipitate of zinc sulphide.

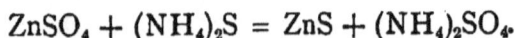

$$ZnSO_4 + (NH_4)_2S = ZnS + (NH_4)_2SO_4.$$

II. Add ammonia drop by drop to another portion. Note that a white precipitate is at first produced; which, however, *readily dissolves as more ammonia is added.*

(This reaction enables us to separate zinc from iron.)

III. **Dry Reaction.**—Heat a little solid zinc sulphate with sodium carbonate upon charcoal in the inner blowpipe-flame. No metallic beads are formed, because zinc oxidizes too easily; but an incrustation of zinc oxide is formed on the charcoal, which appears canary yellow while it is hot, but turns white on cooling. Touch the incrustation with a drop of cobalt nitrate solution, and again heat it in the extreme tip of the flame. The mass becomes green.

Calcium, Ca. (GROUP IV.)

Use calcium chloride, $CaCl_2, 6H_2O$. Dissolve some of the salt in water. Note how quickly and easily it dissolves.

I. Add **ammonium carbonate** (group-reagent) to a portion of the solution. A white precipitate of calcium carbonate is obtained.

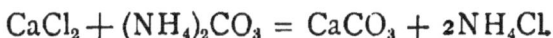

$$CaCl_2 + (NH_4)_2CO_3 = CaCO_3 + 2NH_4Cl.$$

Filter the mixture, and pour a few drops of acetic acid upon the filter. Note effervescence as the precipitate dissolves.

II. Add ammonium oxalate to another portion. A white precipitate of calcium oxalate is obtained.

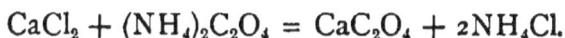

$$CaCl_2 + (NH_4)_2C_2O_4 = CaC_2O_4 + 2NH_4Cl.$$

Pour half this mixture into a second test-tube. To one portion add hydrochloric acid. Note that the precipitate dissolves; to the other add acetic acid. The precipitate does not dissolve.

Dry Reaction.—Dip a clean platinum wire into the calcium chloride solution, and bring it against the edge of a Bunsen flame. Note the reddish colour of the flame.

Potassium, K. (GROUP V.)

Use potassium chloride, KCl. Dissolve a few crystals in a small quantity of water so as to obtain a strong solution.

I. Add platinum chloride (one or two drops) to a small quantity of this solution. A yellow crystalline precipitate is formed, consisting of the compound K_2PtCl_6.

Cautiously add water drop by drop, and gently warm the mixture. Notice that the precipitate dissolves. Hence this test can only be used when the solution is strong. Now add a little strong hydrochloric acid. This causes the reformation of the precipitate. Therefore, before applying this test, it is best to add a drop of strong hydrochloric acid, as this promotes the formation of the precipitate.

II. Add to a second portion of potassium chloride a little strong solution of hydrogen sodium tartrate, a white precipitate of hydrogen potassium tartrate is formed. Add a little water, and note that the precipitate quickly dissolves. Therefore this test also can only be made with strong solutions.

Dry Reactions.—Dip a clean platinum wire into the potassium chloride solution and bring it into the Bunsen flame. A lilac colour is given to the flame. This colour, however, is entirely masked by a yellow colour if any sodium compounds are present, even in the minutest quantities. If the flame be looked at through deep blue glass (or better a glass prism filled with indigo) the colour given by the potassium salt will appear crimson-red in spite of the sodium impurity.

AMMONIUM COMPOUNDS.

These all evolve ammonia, when heated with caustic soda or potash, which is detected by its smell, and by its action on turmeric or litmus paper. See p. 183.

Reactions for Acids.—The four commonest acids are hydrochloric, nitric, carbonic, and sulphuric. The reactions by which these are distinguished have been already described. Chlorides, p. 125; Nitrates, p. 195; Carbonates, p. 230, and sulphates, p. 261.

Method of analysis of a simple mixture.[1]

If the substance given is solid, its general appearance (such as its colour, whether crystalline or not, etc.) should be carefully observed and noted down. Then proceed to make the following tests directly upon the solid.

I. Put a little of the substance into a short, narrow test-tube and heat it. Observe closely what happens.

[1] Containing only chlorides, nitrates, carbonates, or sulphates of ammonium, potassium, calcium, zinc, iron, copper, or lead.

(*a*) If a white sublimate is formed, suspect *ammonium salts.*

Confirm this by heating a little of the salt with caustic soda. Ammonia given off, detected by its smell, and its action on turmeric paper, proves the presence of ammonium salts.

(*b*) If a brown-coloured vapour is given off, suspect a *nitrate*, probably lead nitrate.

Confirm a nitrate by applying the test for nitric acid, given on p. 195.

II. Heat a little of the substance (powdered) on charcoal in the blow-pipe flame (as shown on p. 261).

(*a*) If the substance deflagrates, appearing to go on fire on the charcoal, suspect a *nitrate.*

(*b*) If a bright white residue is left, suspect *calcium* or *zinc* compounds.

Place a portion of the residue upon turmeric paper and moisten with water. If a brown stain is produced, suspect *calcium.*

Moisten a part of the residue on the charcoal with cobalt nitrate, and again heat. If a green residue remains, suspect *zinc.*

(*c*) If metallic beads are formed, suspect *lead.*

Confirm by mixing the substance with sodium carbonate, and heating on charcoal in the inner blow-pipe flame. Malleable bead, which marks paper, confirms *lead.*

III. Heat a little of the substance on platinum wire in a Bunsen flame.

(*a*) If reddish flame, suspect *calcium.* | These must be con-
(*b*) If lilac flame, suspect *potassium.* | firmed later.

IV. Add a little dilute hydrochloric acid to some of the original substance, in a test-tube.

(*a*) If effervescence takes place, suspect a *carbonate.*

Confirm by allowing the gas to enter a tube containing lime water (as shown on p. 227).

V. Gently heat a little of the substance with a few drops of strong sulphuric acid.

(*a*) If acid fumes are evolved, suspect a *chloride* or a *nitrate.*

Dip a glass rod, moistened with silver nitrate, into the mouth of the test-tube.

A white precipitate on the rod indicates *hydrochloric acid.*

Drop a small fragment of copper into the tube. Brown fumes of oxides of nitrogen indicate *nitric acid.*

(Hydrochloric acid may also be confirmed by heating the substance with manganese dioxide and sulphuric acid, when chlorine will be evolved, which is recognized by its colour and its bleaching properties.)

Prepare a solution of the substance.—Place a little of the powder in a test-tube, add water, and gently warm the mixture. If it does not dissolve, add hydrochloric acid,[1] and again boil.

When the solution is obtained, proceed according to the following Table; except that if hydrochloric acid has been used to dissolve the substance the first step is omitted.

---

[1] The only compounds of the prescribed metals and acids (see note, p. 280) which are not soluble in water are the carbonates of lead, copper, iron, and calcium; and the sulphates of lead and calcium. The carbonates dissolve readily in dilute hydrochloric acid, with evolution of carbon dioxide: in the case of lead carbonate, lead chloride is formed, which dissolves on boiling, but crystallizes out again on cooling (see Reaction I., p. 275). Lead sulphate and calcium sulphate are partially dissolved by boiling water.

To the solution in water, add hydrochloric acid so long as a precipitate is formed. Filter the solution.

White precipitate indicates **Lead**.

Confirm by dissolving the precipitate in hot water, and adding potassium chromate; yellow lead chromate is precipitated.

Pass sulphuretted hydrogen into the filtrate. Filter.

Black precipitate indicates **Lead**, or **Copper** (or both).

Boil the precipitate in a few drops of strong nitric acid; dilute with water, add dilute sulphuric acid, and filter.

White residue, $PbSO_4$.

Bluish solution, copper nitrate. Add ammonia; deep blue liquid confirms *copper*.

Boil the filtrate until *quite free from sulphuretted hydrogen*, add a few drops of strong nitric acid, and then add ammonia in excess. Filter.

Brown precipitate indicates **Iron**.

Confirm by dissolving in hydrochloric acid, and adding potassium ferrocyanide. Prussian blue is obtained.

Add ammonium sulphide to the filtrate, and filter.

White precipitate indicates **Zinc**.

Confirm by dissolving in hydrochloric acid, and adding caustic potash. A white precipitate is formed which dissolves in excess of potash. Add sulphuretted hydrogen water, white zinc sulphide is again precipitated.

To the filtrate add ammonium carbonate, and filter.

White precipitate indicates **Calcium**.

Confirm by dissolving in acetic acid, and adding ammonium oxalate; white calcium oxalate is precipitated.

Evaporate the filtrate to dryness, and heat strongly to expel all ammonia salts. Heat until no more fumes are given off. Dissolve residue in small quantity of water, acidify with hydrochloric acid and add platinum chloride. Yellow precipitate indicates **Potassium**.

APPENDIX

FOR the convenience of teachers who may reside in the country, and who have to order chemicals from a distance, the following list has been drawn up. It contains the substances required in the carrying out of the experiments described in this book, with the quantities which it would be suitable to obtain for a small class of from ten to twenty students.

The approximate cost, including the bottles, is £3 5s., but exclusive of platinum wire.

1 Winchester Ammonia (sp. g. 0·880).
    .,   Hydrochloric acid (pure).
    „   Nitric acid (1·42 pure).
    „   Sulphuric acid (pure com.).
1-lb. bottle Alcohol (pure abs.).
    „   Ammonium sulphide.
1 lb. Ammonium chloride (pure).
    „ Lime.
    „ Manganese dioxide.
    „ Marble.
    „ Potassium chlorate.
    „   „ nitrate.
    „ Soda, caustic, sticks.
    „ Sodium carbonate (pure).
    „ Sulphur, roll.
    „ Zinc, granulated.
¼ lb. Alum.
    „ Ammonium carbonate.
    „   „ nitrate.
    „   „ sulphate.
    „ Barium chloride.
    „ Calcium chloride (cryst.).
    „ Copper foil.
    „   „ gauze.
    „   „ sulphate.
    „ Ferrous sulphate.
    „ Ferric oxide.
    „ Glycerin.
    „ Iron pyrites.
    „   „ filings.
    „ Lead acetate.
    „   „ dioxide.
    „   „ nitrate.
    „ Mercury.
    „ Mercuric oxide (red).

¼ lb. Oxalic acid.
    „ Potash, caustic.
    „ Potassium chloride.
    „   „ dichromate.
    „ Red lead.
    „ Sodium nitrate.
    „   „ nitrite.
    „ Sulphur, flowers.
    „ Zinc sulphate.
1 oz. Ammonium thiocyanate.
    „ Aniline blue (or magenta).
    „ Arsenious oxide.
    „ Bromine.
    „ Calcium carbide.
    „ Cobalt chloride.
    „ Copper chloride.
    „   „ nitrate.
    „   „ oxide.
    „ Ferric chloride.
    „ Iodine.
    „ Litmus.
    „ Magnesium carbonate.
    „ Mercuric chloride.
    „ Phosphorus.
    „ Potassium iodide.
    „   „ ferricyanide.
    „   „ ferrocyanide.
    „   „ permanganate.
    „ Sodium.
    „   „ oleate.
    „   „ peroxide.
    „   „ sulphite.
    „ Tartar emetic.
    „ Tartaric acid.
½ oz. Silver nitrate.
¼ oz. Magnesium.
    „ Potassium.
1 drachm Platinum chloride.
1 book Gold leaf.
75 grains Platinum wire (No. 29 standard wire gauge).

INDEX